2D Monoelemental Materials (Xenes) and Related Technologies

Emerging Materials and Technologies

Series Editor
Boris I. Kharissov

Functional Nanomaterials for Regenerative Tissue Medicines
Mariappan Rajan

Uncertainty Quantification of Stochastic Defects in Materials
Liu Chu

Recycling of Plastics, Metals, and Their Composites
R.A. Ilyas, S.M. Sapuan, and Emin Bayraktar

Viral and Antiviral Nanomaterials
Synthesis, Properties, Characterization, and Application
Devarajan Thangadurai, Saher Islam, Charles Oluwaseun Adetunji

Drug Delivery using Nanomaterials
Yasser Shahzad, Syed A.A. Rizvi, Abid Mehmood Yousaf and Talib Hussain

Nanomaterials for Environmental Applications
Mohamed Abou El-Fetouh Barakat and Rajeev Kumar

Nanotechnology for Smart Concrete
Ghasan Fahim Huseien, Nur Hafizah A. Khalid, and Jahangir Mirza

Nanomaterials in the Battle Against Pathogens and Disease Vectors
Kaushik Pal and Tean Zaheer

MXene-Based Photocatalysts: Fabrication and Applications
Zuzeng Qin, Tongming Su, and Hongbing Ji

Advanced Electrochemical Materials in Energy Conversion and Storage
Junbo Hou

Emerging Technologies for Textile Coloration
Mohd Yusuf and Shahid Mohammad

Emerging Pollutant Treatment in Wastewater
S.K. Nataraj

Heterogeneous Catalysis in Organic Transformations
Varun Rawat, Anirban Das, Chandra Mohan Srivastava

2D Monoelemental Materials (Xenes) and Related Technologies: Beyond Graphene
Zongyu Huang, Xiang Qi, Jianxin Zhong

Atomic Force Microscopy for Energy Research
Cai Shen

For more information about this series, please visit:
https://www.routledge.com/Emerging-Materials-and-Technologies/book-series/CRCEMT

2D Monoelemental Materials (Xenes) and Related Technologies

Beyond Graphene

Edited by
Zongyu Huang, Xiang Qi,
and Jianxin Zhong

CRC Press
Taylor & Francis Group
Boca Raton London New York

CRC Press is an imprint of the
Taylor & Francis Group, an **informa** business

First edition published 2022
by CRC Press
6000 Broken Sound Parkway NW, Suite 300, Boca Raton, FL 33487-2742

and by CRC Press
4 Park Square, Milton Park, Abingdon, Oxon, OX14 4RN

CRC Press is an imprint of Taylor & Francis Group, LLC

Library of Congress Cataloging-in-Publication Data
Names: Huang, Zongyu, editor. | Qi, Xiang (Physicist), editor. | Zhong, Jianxin, editor.
Title: 2D monoelemental materials (Xenes) and related technologies: beyond graphene / edited by Zongyu Huang, Xiang Qi, and Jianxin Zhong.
Description: First edition. | Boca Raton: CRC Press, 2022. | Includes bibliographical references and index. | Summary: "This book describes the structure, properties, and applications of monoelemental 2D materials (Xenes) by classification and section. The first section covers the structure and classification of single-element 2D materials, according to the different main groups of monoelemental materials of different components of the properties and applications of detailed description. The second discusses the structure, properties, and applications of advanced 2D Xenes materials. Aimed at researchers and advanced students in materials science and engineering, this book offers a broad view of current knowledge in the emerging and promising field of 2D monoelemental materials"–Provided by publisher.
Identifiers: LCCN 2021049217 (print) | LCCN 2021049218 (ebook) | ISBN 9781032074788 (hardback) | ISBN 9781032074801 (paperback) | ISBN 9781003207122 (ebook)
Subjects: LCSH: Two-dimensional materials. | Layer structure (Solids)
Classification: LCC TA418.9.T96 A122 2022 (print) | LCC TA418.9.T96 (ebook) | DDC 620.1/12–dc23/eng/20211117
LC record available at https://lccn.loc.gov/2021049217
LC ebook record available at https://lccn.loc.gov/2021049218

ISBN: 978-1-032-07478-8 (hbk)
ISBN: 978-1-032-07480-1 (pbk)
ISBN: 978-1-003-20712-2 (ebk)

DOI: 10.1201/9781003207122

Typeset in Times
by Deanta Global Publishing Services, Chennai, India

Contents

Preface

Since the successful discovery of graphene, two-dimensional (2D) materials have become a hot topic for researchers. Due to the reduction in size, 2D materials show many novel properties such as high strength, high mobility, and high flexibility. Therefore, they have potential applications in electron, photoelectron, catalysis, and energy storage.

2D materials include layered and non-layered structures. Layered 2D materials are held together by covalent bonds, and the layers are connected by weak van der Waals forces (vdWs), while all the atoms of non-layered 2D materials are connected by chemical bonds. Monoelement 2D materials with similar structures to graphene have been widely studied for their properties and applications such as topology, lithium battery storage, and high electrical conductivity. Transitional metal dichalcogenides (TMDs) due to the rich chemical composition and structural phase diversity, from semiconductors to metals, show rich photoelectron properties, superconductivity, and charge density wave, etc. 2D metal carbides and nitrides (MXenes) materials have shown great potential in the fields of sensors, photothermal conversion, and field-effect transistors. In addition, the 2D hexagonal boron nitride (h-BN) has excellent lubricity, high temperature resistance, and chemical corrosion resistance and is often used as a good base material. Besides the typical 2D layered materials mentioned above, the properties of 2D perovskite materials, such as a high absorption coefficient, long-distance carrier diffusion, and the high porosity and large inner specific surface area of 2D metal-organic frames, mean these non-layered materials have potential applications in many aspects. Starting from 2D materials with basic structures, more emerging 2D materials with rich properties will be constructed by conventional and effective means such as constructing heterojunctions, generating defects, and using external action, including forces, heat, light, and electromagnetism.

This book divides 2D materials into two categories: layered and non-layered, with the 2D materials of monoelemental, the transitional metal dichalcogenides, the metal carbides and nitrides (MXenes), and the hexagonal boron nitride (h-BN) and related materials; the non-layered 2D materials are represented by the 2D perovskite materials and 2D metal organic framework. In addition, the structure, properties, and applications of newly emerging 2D materials, which are constructed by heterogeneous junctions, produced by defects, and regulated by external fields, are described in detail as separate chapters.

Editors

Zongyu Huang is Professor and Director for the Department of Physics, School of Physics and Optoelectronics, Xiangtan University, China. She earned her B.S. degree in Physics from Hubei University, China, in 2004, her M.S. degree in Condensed Matter Physics from Wuhan University, China, in 2006, and her Ph.D. degree in Physics from Xiangtan University, China, in 2014. Her research interest is in physical properties and applications of two-dimensional (2D) materials via first principles calculation methods, and advanced synthesis and characterization methods. Her work is documented in over 50 peer-reviewed publications in international journals, such as *Advanced Energy Materials* and *Nano Today*.

Xiang Qi is Professor and Vice-Director of the Hunan Key Laboratory of Micro-Nano Energy Materials and Devices at Xiangtan University, China. He earned his B.S. and Ph.D. degrees from Wuhan University, China, in 2004 and 2009, respectively. Previously he was Visiting Professor at Monash University, Australia (2016–2017), and Visiting Student at Nanyang Technological University, Singapore (2007). His research interests are focused on the field of 2D material and its applications, with particular emphasis on monoelement 2D materials and their-derived structures in optoelectronic, clean renewable energy and environmental science. Currently, he has published over 60 papers with total citations over 6370 and an H-index of 42 based on the data from ISI Web of Science. Twenty of his publications have been cited more than 100 times.

Jianxin Zhong is Professor in Condensed Matter and Materials Physics, Dean of the School of Physics and Optoelectronics, and Director of the Hunan Key Laboratory of Micro-Nano Energy Materials and Devices at Xiangtan University, China. He earned his B.S. degree in Physics from Xiangtan University, China, in 1985, his M.S. degree in Theoretical Physics from Xiangtan University and Hunan University, China, in 1989, and his Ph.D. degree in Solid State Physics from the University of Cergy-Pontoise, France, in 1995. Previously, he was Visiting Professor at the Max Planck Institute for Dynamics and Self-Organization and the Chemnitz University of Technology, Germany (1997–1998), Research Scientist (1998–2007) and Distinguished Visiting Scientist (2008–2014) at Oak Ridge National Laboratory and the University of Tennessee, Knoxville. His research interest is in synthesis and properties of functional nanomaterials, focusing on 2D materials, topological insulators, and semiconductor nanomaterials for energy applications. Research tools include first principles methods, molecular dynamic simulations, and advanced synthesis and characterization methods. His work is documented in over 300 peer-reviewed publications in international journals. He has received numerous national and international awards of excellence over the years, including a 2007 MICRO/NANO 25 Award from R&D 100 Magazine, Progress in Science and Technology Award by the Ministry of Education of China, Hunan Provincial Natural Science Award, National Model Teacher Award, and National Distinguished Teacher Award.

Contributors

Yexin Feng
Hunan Provincial Key Laboratory of
Low Dimensional Structural Physics
& Devices
School of Physics and Electronics
Hunan University
Changsha, China

Gencai Guo
School of Physics and Optoelectronics
Xiangtan University
Xiangtan, China

And

College of Materials Science and
Engineering
Beijing University of Technology
Beijing, China

Rong Hu
Hunan Key Laboratory for Micro-
Nano Energy Materials and Devices
School of Physics and Optoelectronics
Xiangtan University
Hunan, People's Republic of China

Zongyu Huang
Hunan Key Laboratory for Micro-
Nano Energy Materials and Devices

And

School of Physics and Optoelectronics
Xiangtan University
Xiangtan, China

Jun Li
Hunan Key Laboratory of Micro-Nano
Energy Materials and Devices
Laboratory for Quantum
Engineering and Micro-Nano Energy
Technology

And

School of Physics and Optoelectronics
Xiangtan University
Hunan, People's Republic of China

Zhenqing Li
Shenzhen Geim Graphene Center
Tsinghua-Berkeley Shenzhen Institute
and Tsinghua Shenzhen International
Graduate School
Tsinghua University
Shenzhen, China

Yujie Liao
School of Physics and Optoelectronics
Xiangtan University
Xiangtan, China

Huating Liu
Hunan Key Laboratory for Micro-
Nano Energy Materials and Devices

And

School of Physics and Optoelectronics
Xiangtan University
Hunan, People's Republic of China

Yundan Liu
School of Physics and Optoelectronics
Xiangtan University
Xiangtan, China

Siwei Luo
School of Physics and Optoelectronics
Xiangtan University
Xiangtan, China

Dan Mu
School of Physics and Optoelectronics
Xiangtan University
Xiangtan, China

Xiang Qi
Hunan Key Laboratory for Micro-
Nano Energy Materials and Devices

And

School of Physics and Optoelectronics
Xiangtan University
Hunan, People's Republic of China

Hui Qiao
Hunan Key Laboratory for Micro-
Nano Energy Materials and Devices

And

School of Physics and Optoelectronics
Xiangtan University
Hunan, People's Republic of China

Kai Wang
Hunan Key Laboratory of Micro-Nano
Energy Materials and Devices
Laboratory for Quantum Engineering
and Micro-Nano Energy Technology

And

School of Physics and Optoelectronics
Xiangtan University
Hunan, People's Republic of China

Xiong-Xiong Xue
School of Physics and Optoelectronics
Xiangtan University
Xiangtan, China

Sifan Zhang
School of Physics and Optoelectronics
Xiangtan University
Xiangtan, China

Jianxin Zhong
Hunan Key Laboratory for Micro-
Nano Energy Materials and Devices

And

School of Physics and Optoelectronics
Xiangtan University
Hunan, People's Republic of China

Jincheng Zhuang
School of Physics
Beihang University
Beijing, China

1 Structure and Classification of 2D Monoelemental Materials (Xenes)

Sifan Zhang and Zhenqing Li

CONTENTS

1.1 INTRODUCTION

In 2004, when monolayer graphene was experimentally prepared from graphite by a mechanical stripping method, the era of monolayer two-dimensional (2D) materials was officially inaugurated.[1] Researchers found that graphene has excellent properties such as high mobility and excellent electrical and thermal conductivity.[2–4] It is worth mentioning that a large number of theoretical and experimental studies have shown that the electrons in graphene have linear Dirac fermion properties, which gives graphene even more novel physical properties.[5] These excellent properties are due to the unique two-dimensional structure and orbital hybridization of graphene instead of the three-dimensional structure.[6] However, graphene has been found to have a zero band gap, which limits its application in optoelectronics.[7] To keep its advantages and overcome its disadvantages, other monoelemental 2D materials have been studied gradually.[1]

DOI: 10.1201/9781003207122-1

2D Xenes mainly refer to 2D monolayers of elemental materials composed of group IIIA, group IVA, group VA, and group VIA.[1] Up to now, 15 kinds of monoelemental 2D layered materials have been experimentally and theoretically confirmed. The general classification of emerging 2D Xenes and their corresponding synthesis methods are presented in Figure 1.1[8, 9]. Except for aluminene and indiene which only exist as a theoretical prediction, all the other 13 elemental 2D forms have been successfully synthetized or fabricated.[10] They include borophene,[11] aluminene,[12] gallenene, and indiene in group IIIA,[13, 14] silicene,[15] germanene,[16] stanene and plumbene in group IVA,[17, 18] phosphorene,[19] arsenene,[20] antimonene and bismuthene in group VA,[21, 22] and selenene and tellurene in group VIA.[1] And they have many excellent properties and applications, including electronics, spintronics, optoelectronics, energy, and so on, as shown in Figure 1.1.

The structure of a substance determines its properties. Symmetry, allotropes, and orbital hybridization has vital effects on the structure. Here, we illustrate the importance of symmetry, phase, and orbital hybridization of the structure by taking some 2D Xenes as an example.[23] At the same time, the structure of 2D Xenes are systematically classified and described.[1] This paper is helpful for readers to better understand the structure of monoelemental 2D materials and the effect of structure on the application of properties, and provides some theoretical guidance for their further research and development.

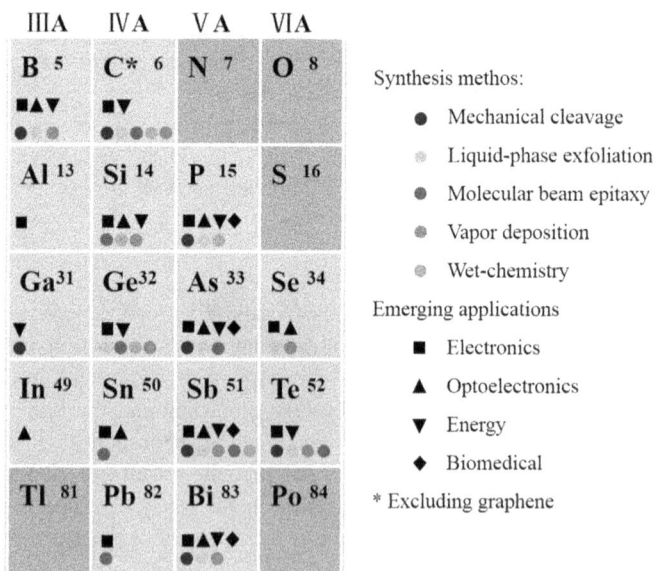

FIGURE 1.1 2D Xenes materials with their known synthesis methods and applications. Elements shown in nattier blue and light sea green denote that elements have been predicted and have not been predicted from Xenes, respectively.

1.2 THE FACTORS OF 2D XENES' STRUCTURES

1.2.1 SYMMETRY

Symmetry is the property of a crystal that has the same atomic structure in corresponding directions, and in symmetric mirror relations along these directions, and that is similar in physical and crystallographic terms in two or more directions. Symmetries in most crystals mainly include spatial translation, axis rotation, center inversion, mirror inversion, and so on. Among them, spatial translation is the simplest symmetry of crystal materials. This symmetry is referred to as long-range ordering, which is mainly described by hexahedral cell units surrounded by three vectors on the atomic and molecular levels. Symmetry breaking refers to the phenomenon where asymmetry factors occur and the degree of symmetry spontaneously decreases. Symmetry is ubiquitous in the system of all scales. Symmetry breaking is the way of difference of things and any symmetry must have symmetry breaking. Breaking symmetry and breaking defects are also of great significance for structural exploration.

Among the physical properties of 2D Xenes, i.e., structural properties, are some of the most unique properties compared to other non-2D materials. 2D Xenes have ultrahigh surface area to volume. Different from three-dimensional materials, two-dimensional materials are inherently constrained by the degree of freedom in the direction perpendicular to the material. The buckled structures of 2D Xenes make their atoms stronger in overlapping, leading to a hybridized sp^2-sp^3 configuration with a lower symmetry and stability. Breaking symmetry leads to novel physical properties. Xenes can be functionalized by substrate, chemical adsorption, defects, charge doping, external electric field, periodic potential, in-plane uniaxial and biaxial stress, and out-of-plane long-range structural deformation, to name a few.[24] The electronic properties, including the magnitude of the energy gap, can further be tuned and controlled by external means.

1.2.2 ALLOTROPES

An allotrope is a substance composed of the same single chemical element but with different properties, because the same element atoms are arranged in different ways resulting in different physical and chemical properties of the structure. The properties of 2D materials are dictated by their allotropes, how atoms are arranged in the lattice. There are many allotropes in nature, and they have completely different physical properties. For example, graphite and diamond are typical allotropes of carbon.

The 2D materials have three kinds of crystal bonding structures, which are simplified into the lattice diagram shown in Figure 1.2 (a), which is mainly divided into three modes: hexagonal lattice, square lattice, and honeycomb lattice. The bonds are the longest in the hexagonal lattice, intermediate in the square lattice, and shortest in the honeycomb lattice, and the bond angles are 60o (hex),

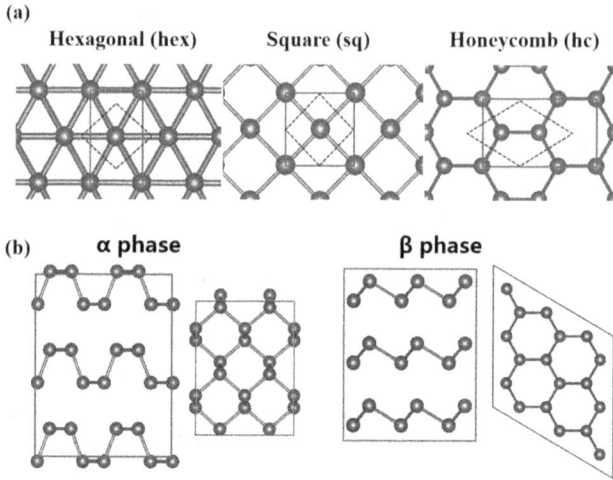

(a)

Hexagonal (hex)   Square (sq)   Honeycomb (hc)

(b)   α phase   β phase

FIGURE 1.2   (a) Schematic representations of the 2D lattices. The solid line indicates the computational unit cells and dashed lines the Wigner-Seitz cells. (b) Structures for phosphorus (α and β phase).

90∘ (sq), and 120∘ (hc). In 2D, the number of nearest neighbors is six for hexagonal, four for square, and three for honeycomb 2D lattices. Most metals are stable in the hexagonal and honeycomb geometries, but unstable in the square-lattice geometry. Most allotropes occur on nonmetallic elements, so allotropes are easy to appear in 2D Xenes' materials. For example, black phosphorus presents different phases at different temperatures during the preparation process. At high temperatures, as shown in Figure 1.2 (b), in each single atom layer, the phosphorus atom forms a unique puckered structure (α phase) because two covalent bonds lie parallel with the atomic plane while the third one is nearly perpendicular to that, which plays an important role in bridge connection of phosphorus from upper and lower layers. At low temperatures, black phosphorus transforms into a semimetallic β phase as seen in Figure 1.2 (b). It has been theoretically confirmed that two types of buckling, low-buckled and high-buckled, exist among many hexagonal 2D Xenes.

The effect of phase on structure has attracted scientists to further investigate the allotropes of monoelemental materials and their excellent properties have also attracted many scientists to conduct structural prediction research on 2D Xenes' materials. For example, He et al. used the RG2 method to search and predict the two-dimensional graphene allotropes, and obtained 1116 new allotropes.[25] Meanwhile, the electronic properties of these two-dimensional graphene allotropes were calculated by using the tight-binding method, and it was found that there were 685 metal structures, 241 semiconductor structures, and 190 semimetal structures.[26] The greatly enriches the properties and applications of two-dimensional graphene allotropes.

1.2.3 ORBITAL HYBRIDIZATION

Orbital hybridization has an important influence on the properties and applications of structures. The bonding of atoms is very important because crystals are formed by interactions between atoms. In this process, like all spontaneous processes in nature, they tend to have the lowest energy of the whole system. This is because the lower the energy, the more stable the system. To achieve this minimum energy state, the bonding atomic orbitals have overlap as much as possible and increase the number of bonds, that is, when the atoms interact with each other; only through these two aspects can the energy minimum be achieved. So orbital hybridization is essential. At the same time, after hybridization of orbitals, electron clouds are more concentrated in the direction of bonding and more effectively overlap bonding. Compared with pure s, p, and d orbitals, the bonding energy can be exerted more effectively and the crystal structure is more stable. There are currently eight basic types of orbital hybridization as shown in Table 1.1.

Different structures have different orbital hybridization states. For example, the outermost electron configuration of a carbon atom is $[He]2s^22p^2$, which can form a linear sp bond (alkyne bond), planar sp^2 bond (ene bond), and tetrahedral sp^3 bond (alkane bond). Therefore, 3D block and 2D layered materials have different orbital hybridization states. Taking 2D layered graphene as an example, a two-dimensional hexagonal crystal structure is formed mainly by sp^2 hybridization between C atoms to form covalent bonds (as shown in Figure 1.2 (a)). The three sp^2 hybridized orbitals have the same shape and different spatial orientations. The three one-electron sp^2 orbitals on each carbon atom form σ bonds with each of the three adjacent carbon atoms. The formation of a hexagonal honeycomb-like planar layered structure takes place. Each carbon atom also has a single electric sp orbital that is not involved in the hybridization. These p orbitals are perpendicular to the plane of the carbon sp2 hybridization orbital and are parallel to each other to form delocalized large π bonds, making graphene have very good electrical properties.

TABLE 1.1
The Type of Orbital Hybridization

Central Atomic Hybrid Type	Atomic Configuration Type
sp	Straight line
pd	Straight line
sp^2	Plane triangle
sp^3	Tetrahedron
sp^3d	Tripartite biconical
$sp^3d_{z2}d_{x2-y2}$ (sp^3d^2)	Octahedron
$sp^3d_{xz}d_{xy}d_{yz}$ (sp^3d^3)	Pentagonal biconical
$sp_xp_yd_{x2-y2}$ (sp^2d)	Flat square

The 2D Xenes' materials are mainly distributed in three, four, five, and six main groups and the electron configuration in the outermost atom is mainly s and p orbitals. In contrast to graphene, 2D Xenes are not completely planar and feature a low-buckled geometry with two sublattices displaced vertically as a result of the interplay between sp^2 and sp^3 orbital hybridization. Therefore, this paper mainly analyzes the hybridization state of orbitals between s and p.[23]

1.3 THE CLASSIFICATION OF 2D XENES

1.3.1 GROUP III

Group IIIA in the periodic table, in column 13, has the shell electron structure of $[He]2s^22p^1$. Among group IIIA 2D Xenes, borophene (B), aluminene (Al), gallenen (Ga), and indiene (In) have been studied in the recent literature. There are many allotropes of these group IIIA elements that have been theoretically predicted. This complexity could possibly be related to the lower number of three valence electrons in group IIIA elements, resulting in their unique electronic band structures. These structures are shown in Figure 1.3.

Boron is the only nonmetallic element in group IIIA. 2D boron (borophene) properties have been probed theoretically by first principles, and 2D boron

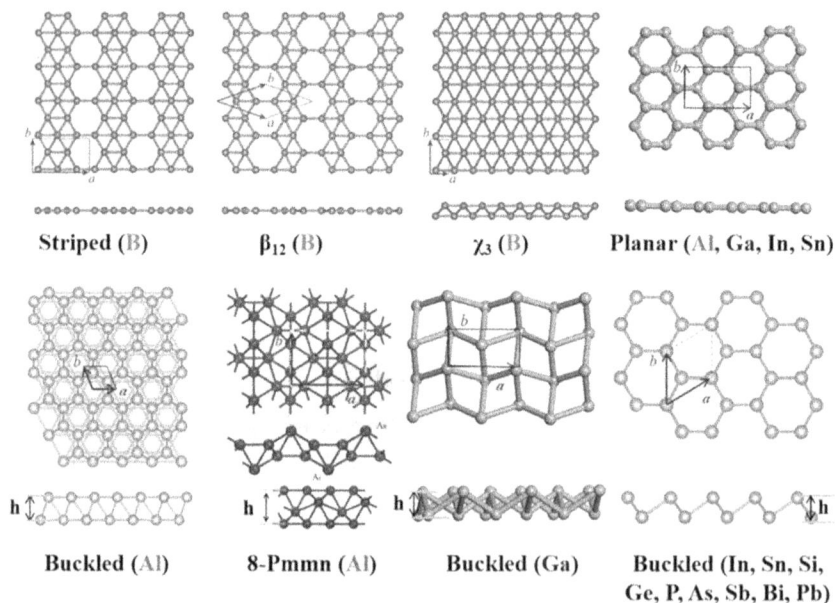

Striped (B) β_{12} (B) χ_3 (B)   Planar (Al, Ga, In, Sn)

Buckled (Al)   8-Pmnn (Al)   Buckled (Ga)   Buckled (In, Sn, Si, Ge, P, As, Sb, Bi, Pb)

FIGURE 1.3 The allotrope structure of various elements atoms. Image for "8-Pmnn" (Group IIIA) is reproduced from Ref. 31 with permission. Copyright 2019, Elsevier B.V.

allotropes with a large number of possible low-energy structures have been predicted.[27, 28] These allotropes differ from their parent structure, the buckled triangular boron, via concentration of vacant B sites as shown in Figure 1.3. These allotropes are not all in a low-energy stable state.[29] For example, free-standing triangular boron and χ_3 have been predicted to be dynamically unstable, with a stability order of $\chi_3 > \beta_{12} >$ striped triangular boron.[30] It has been shown that 2D boron is a metal, in contrast to its 3D crystal being a semiconductor. These borophene allotropes have been confirmed by experimental studies, and they could be obtained by mechanical cleavage, liquid-phase exfoliation, and vapor deposition ways as shown in Figure 1.1.[1]

Aluminum (aluminene), gallium (gallenene), and indium (indiene) also have shown different structures, as can be seen in Figure 1.1.[1] Aluminene has been studied and found in three different allotropes (planar, buckled, and curved).[31] We would like to note that these allotropes are stable crystal structures, and exhibit metallic and semiconductor characteristics, respectively. Gallenene has also been reported and shows two stable 2D configurations (planar and curved). Gallenene could be used by a simple stripping technique to peel a thinner 2D gallium layer from the molten phase of Ga.[32] Indium monolayers also have three different allotrope forms (planar, puckered, and buckled).[33] And the study shows that the planar and buckled allotropes are stable and show metallic and semiconducting behavior, respectively. The aluminene and indiene are currently only theoretically predicted and have not been experimentally realized.[1]

1.3.2 GROUP IV

The group IVA 2D Xenes includes silicene (Si), germanene (Ge), stanene (Sn), and plumbene (Pb).[33–35] These four 2D Xenes possess similar electronic properties as graphene because the shell electron structure of these 2D Xenes is the same as graphene. But in obtaining hybridization of silicene, gallenene, and stanine, they do not tend to undergo sp^2 hybridization. Silicene is the first and most widely investigated of group IVA 2D Xenes. Contrary to graphene with sp^2 hybridization, free-standing silicene adopts a buckled hexagonal honeycomb structure.[36] The buckled Si atoms make themselves closer to enable a stronger overlapping, resulting in a hybridized sp^2-sp^3 configuration with lower symmetry.[23] Germanene with mixed sp^2-sp^3 hybridization leads to the buckled honeycomb structure.[37] This leads to all the 2D Xenes except graphene being not exactly planar, which is called buckled structure as shown in Figure 1.3, which has a mixed degree of sp^2-sp^3 character. The silicene and germanene are energetically more stable in the low-buckled phase, while stanene and plumbene prefer the high-buckled phase to become more stable. In comparison to graphene, the larger distance of atomic bonding of germanene results in stronger spin-orbit interaction. Stanene has been predicted to have a buckled honeycomb lattice structure.[38]

In fact, the reason that 2D Xenes are buckled rather than planar is that the honeycomb planar lattice is not stable. The π-π overlap of four Xenes is much weaker than in the case of graphene due to the larger atomic radius of Si, Ge, Sn, and Pb atoms. The weak π-π overlap caused by the repulsive Coulomb potential between cores results in the instability of the planar honeycomb lattice. To reduce these repulsive atoms, the atoms of the 2D honeycomb lattice are arranged off their plane to form two sublattices with opposite outward directions. Thus, the atomic structure becomes buckled or puckered on the atomic scale.

Moreover, the change in buckling height causes a drastic change in the electronic properties in the band structure. The study of the structural and electronic properties of IVA group elements is shown in Table 1.2. Both the lattice constant (a) and bond length (l) increases from C to Pb. The pristine hexagonal monolayer Xenes show buckled structure upon relaxation, except for graphene. The buckling h increases linearly from graphene to plumbene, which is consistent with the effect of increasing atomic radius on the degree of 2D Xenes buckled. However, the bandgap energy (Eg) decreases down along the group, ranging from a small value of 1.55 meV (silicene) to as large as 440 meV (plumbene). The results show that these four Xenes make up the defect of graphene zero band gap, and they also have Dirac cones.[39] Thus, the Dirac cones are not limited to graphene, but they turn out to be more prevalent.

1.3.3 Group V

The group VA Xenes are phosphorene (P), arsenene (As), antimonene (Sb), and bismuthene (Bi).[1] There are a variety of allotropes in group VA elements. In nature, phosphorus has white, red, black phosphorus and other several amorphous forms.[40, 41] Arsenic has three common allotropes, which are metallic gray, yellow, and black arsenic.[42] Antimony has three allotropes, which are gray antimony, black antimony, and explosive antimony.[1] And bismuth has one stable structure, which is gray arsenic. Among them, the most stable structures are black phosphorus, gray arsenic, gray antimony, and gray arsenic. There are two phases that have been isolated in natural layered crystals, α phase and β phase, as shown in Figure 1.2. With the exploration of the five groups of phosphorene allotropes by theoretical calculation, more phase structures have been found. Zhang et al. systematically

---

TABLE 1.2

Optimized 2D Lattice Constant a, Buckling Height h, Bond Angle α, and Band Gap with SOC for Buckled Structures of Group IV Xenes.

Element	a (Å)	l (Å)	h (Å)	E^{SO}_g (meV)
Graphene (C)	2.46[35]	1.42[35]	0.00[35]	0.00[48]
Silicene (Si)	3.87[49]	2.28[49]	0.45[49]	1.55[50]
Germanene (Ge)	4.06[51]	2.44[51]	0.67[51]	23.9[52]
Stanene (Sn)	4.55[53]	2.70[53]	0.81[53]	74.0[53]
Plumbene (Pb)	4.93[54]	3.00[54]	0.99[54]	440[54]

proposed nine-phase structures, including five typical honeycomb (α, β, γ, δ, ε) and four non-honeycomb (ζ, η, θ, ι) structures as shown in Figure 1.4.[4, 3]

Among these structures, phosphorene is the most stable in phase α, which is black phosphorus with puckered structures as shown in Figure 1.2 (b). The phosphorus atom has five valence electrons from the shell electron structure [Ne]$3s^23p^3$. The phosphorus atoms resemble a puckered honeycomb structure by sp3 orbital hybridization, where each phosphorus atom is covalently bonded with two neighboring intraplane atoms and bonded with an adjacent interplane by its p orbitals, which hybridize to form three covalent bonds with neighboring atoms. There are two covalent bonds that lie parallel to the atomic plane. And three covalent bonds are nearly perpendicular to the plane, which plays an important role in the bridge connection of two phosphorus atoms layers. So, it forms a α-phase puckered structure as shown in Figure 1.2 (b). Black phosphorene has direct bandgap, which is according to both experiments and theoretical results.[40] And black phosphorene acts like a bridge between graphene and TMDs. Blue phosphorene with the modified structure of zigzag puckering or buckling is shown in Figure 1.2 (b) β-phase.[44]

Different from black phosphorene, the β-phase is the most stable for arsenene, antimonene, and bismuthene monolayer allotropes. Actually, the natural layered crystals of As, Sb, and Bi have a β phase. It is obvious that the calculation results

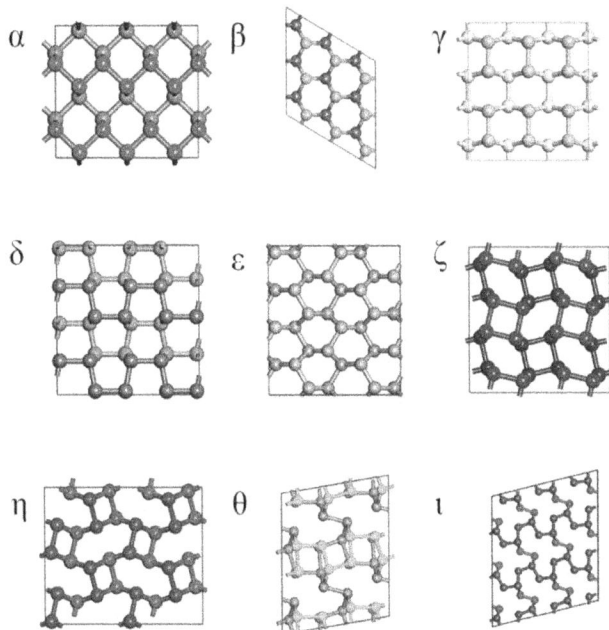

FIGURE 1.4 There are five typical honeycomb (α, β, γ, δ, ε) and four non-honeycomb (ζ, η, θ, ι) structures of the group VA allotropes. Reproduced from Ref. 43 with permission. Copyright 2016, Wiley.

accord with the experiment and the natural situation. In monolayer arsenene, each As atom is bonded with three adjacent atoms and then built as a stable buckled honeycomb structure layer by layer.[42] Antimonene with α and β allotropes was presented as stable.[45]

1.3.4 GROUP VI

The group VIA elements have 3D bulk structures in nature. Compared with other Xenes of 2D Xenes, the group VIA of single-element materials has only been studied in recent years. 2D group VIA Xenes are mainly composed of selenium (Se) and tellurium (Te).[46, 47] The selenene could be obtained by physical vapor deposition and liquid-phase ways. And the tellurene could be obtained by mechanical cleavage, liquid-phase exfoliation, molecular beam epitaxy, and vapor deposition ways as shown in Figure 1.1. The 2D group VIA Xenes also have allotropes with different phases.

The monolayer structure for selenene and tellurene has been predicted by first principles calculations. As shown in Figure 1.5, Se atoms are arranged in a buckled square lattice within this structure. The atomic structures of selenene are composed of atomic rings or 1D helical chains. Two stable phases of selenene, which are α-Se and β-Se with deformed four-atom and six-atom rings, are identified. The structure is a bit similar to those of group IVA 2D Xenes' buckled structure. Tellurene crystallizes in a triangular spiral structure composed of chains of Te atoms that are stacked together in a hexagonal array by van der Waals (vdWs) forces. In this structure, the Te atom is covalently bonded with the adjacent Te atoms on the same chain as shown in Figure 1.5. And elemental Te tends to form

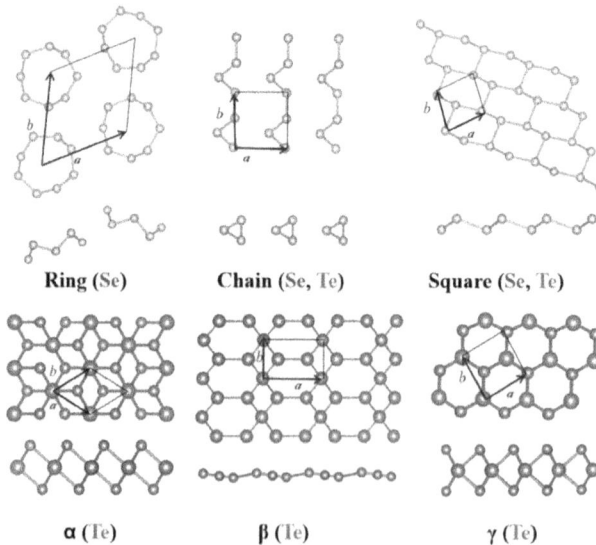

Ring (Se)	**Chain (Se, Te)**	**Square (Se, Te)**
α (Te)	**β (Te)**	**γ (Te)**

FIGURE 1.5   Top and side view of 2D selenene and tellurene allotropes structures.

1D helical structures because Te always forms a chiral crystal, which is the most stable configuration of these allotropes.

1.4 SUMMARY AND PERSPECTIVES

This paper systematically classifies and summarizes the structures of the 2D Xenes in terms of their symmetry, allotrope, and orbital hybridization. We present our insights with the aim of better analyzing the effects of structure on properties. The results show that the diversity of 2D Xenes depends on the change of structures. All 2D Xenes have a buckled structure. The buckled height of group IVA increases with the increase of atomic number, and the buckled structures of group IIIA and group VIA are also affected by the structural changes. Among them, the allotropes of group VA have the most allotropes.

As the extensive investigations on 2D Xenes' materials with atomic-scale thickness show, their air stability and integration capability issues are exposed against their practical applications. 2D Xenes still have some challenges in the experimental preparation process. Firstly, the prediction and experimental preparation of some 2D Xenes remain some of the most challenging problems, such as preparation of Tl, N, O, S, and Po elements in the atomic composition of 2D Xenes. Secondly, the preparation of large-scale high-quality 2D Xenes remain a technical challenge, and how to improve production efficiency and material properties is also an important research direction. In addition, 2D Xenes will have a wide range of applications in electronics, optoelectronics, energy, and biomedical fields. So it is very necessary to further expand the application of 2D Xenes to other new application fields.

REFERENCES

1. Novoselov, K. S. & Geim, A. K. et al., Electric field effect in atomically thin carbon films. *Science*, 306, 5696, 2004.
2. Changgu, Lee, Xiaoding, Wei, Kysar Jeffrey, W. & James, Hone, Measurement of the elastic properties and intrinsic strength of monolayer graphene. *Science*, 321, 5887, 2008.
3. Butler, Sheneve Z., et al., Progress, challenges, and opportunities in two-dimensional materials beyond graphene. *ACS Nano*, 7, 4, 2013.
4. Bolotin, K. I. et al., Ultrahigh electron mobility in suspended graphene. *Solid State Communications*, 146, 9–10, 2008.
5. Katsnelson, M. I., Graphene: carbon in two dimensions. *Materials Today*, 10, 1–2, 2007.
6. Latil, S. & Henrard, L., Charge carriers in few-layer graphene films. *Physical Review Letters*, 97, 3, 2006.
7. Gui, G., Li, J. & Zhong, J., Band structure engineering of graphene by strain: first-principles calculations. *Physical Review B*, 78, 7, 2008.
8. Feng, B. et al., Experimental realization of two-dimensional boron sheets. *Nature Chemistry*, 8, 6, 2016.
9. Wang, Yao et al., Two-dimensional ferroelectricity and switchable spin-textures in ultra-thin elemental Te multilayers. *Materials Horizons*, 5, 3, 2018.

10. Lin, Ren Jie et al., Achromatic metalens array for full-colour light field imaging. *Nature Nanotechnology*, 14, 3, 2019.
11. Huang, Y., Shirodkar, S. N. & Yakobson, B. I., Two-dimensional boron polymorphs for visible range plasmonics: a first-principles exploration. *Journal of the American Chemical Society*, 139, 47, 2017.
12. Kamal, C., Chakrabarti, A. & Ezawa, M., Aluminene as highly hole-doped graphene. *New Journal of Physics*, 17, 8, 2015.
13. Tao, M.-L. et al., Gallenene epitaxially grown on Si(1 1 1). *2D Materials*, 5, 3, 2018.
14. Shengli, D. et al., Atomically thin arsenene and antimonene: semimetal–semiconductor and indirect–direct band-gap transitions. *Angewandte Chemie - International Edition*, 127, 10, 2015.
15. Vogt, P. et al., Silicene: compelling experimental evidence for graphenelike two-dimensional silicon. *Physical Review Letters*, 108, 15, 2012.
16. Ianco, E. B. et al., Stability and exfoliation of germanane: a germanium graphane analogue. *Acs Nano*, 7, 5, 2013.
17. Yuhara, J., He, B., Matsunami, N., Nakatake, M. & Lay, G. L., Graphene's latest cousin: plumbene epitaxial growth on a 'nano water-cube'. *Advanced Materials*, 31, 27, 2019.
18. Zhu, F. F. et al., Epitaxial growth of two-dimensional stanene. *Nature Materials*, 14, 10, 2015.
19. Qiao, J., Kong, X., Hu, Z. X., Feng, Y. & Ji, W., High-mobility transport anisotropy and linear dichroism in few-layer black phosphorus. *Nature Communications*, 5, 1, 2014.
20. Zhang, S., Yan, Z., Li, Y., Chen, Z. & Zeng, H., Atomically thin arsenene and antimonene: semimetal-semiconductor and indirect-direct band-gap transitions. *Angewandte Chemie - International Edition*, 54, 10, 2015.
21. Reis, F. et al., Bismuthene on a SiC substrate: a candidate for a high-temperature quantum spin Hall material. *Science*, 357, 6348, 2017.
22. Fei, R. & Yang, L., Strain-engineering the anisotropic electrical conductance of few-layer black phosphorus. *Nano Letters*, 14, 5, 2014.
23. Molle, A. et al., Buckled two-dimensional Xene sheets. *Nature Materials*, 16, 2, 2017.
24. Zhang, L. et al., Recent advances in hybridization, doping, and functionalization of 2D xenes. *Advanced Functional Materials*, 31, 1, 2020.
25. Shi, X., He, C., Pickard, C. J., Chao, T. & Zhong, J., Stochastic generation of complex crystal structures combining group and graph theory with application to carbon. *Physical Review B*, 97, 1, 2018.
26. He, C. et al., Complex low energy tetrahedral polymorphs of group IV elements from first-principles. *Physical Review Letters*, 121, 17, 2018.
27. Sun, X. et al., Two-dimensional boron crystals: structural stability, tunable properties, fabrications and applications. *Advanced Functional Materials*, 27, 19, 2016.
28. Ji, X. et al., A novel top-down synthesis of ultrathin 2D boron nanosheets for multimodal imaging-guided cancer therapy. *Advanced Materials*, 30, 36, 2018.
29. Li, W. L. et al., Recent progress on the investigations of boron clusters and boron-based materials (I): borophene. *SCIENTIA SINICA Chimica*, 48, 2, 2018.
30. Peng, B. et al., The electronic, optical, and thermodynamic properties of borophene from first-principles calculations. *Journal of Materials Chemistry C*, 4, 2016.
31. Yuan, J., Yu, N., Xue, K. & Miao, X., Stability, electronic and thermodynamic properties of aluminene from first-principles calculations. *Applied Surface Ence*, 409, Jul. 1, 2017.
32. Kochat, V. et al., Atomically thin gallium layers from solid-melt exfoliation. *Science Advances*, 4, 3, 2018.

33. Kecik, D., Durgun, E. & Ciraci, S., Optical properties of single-layer and bilayer arsenene phases. *Physical Review B*, 94, 20, 2016.
34. Krawiec, M., Functionalization of group-14 two-dimensional materials. *Journal of Physics Condensed Matter*, 30, 23, 2018.
35. Balendhran, S. et al., Elemental analogues of graphene: silicene, germanene, stanene, and phosphorene. *Small*, 6, 11, 2015.
36. Ni, Z. et al., Tunable bandgap in silicene and germanene. *Nano Letters*, 12, 1, 2011.
37. Acun, A. et al., Germanene: the germanium analogue of graphene. *Journal of Physics Condensed Matter An Institute of Physics Journal*, 27, 44, 2015.
38. Zhou, H., Cai, Y., Gang, Z. & Zhang, Y. W., Quantum thermal transport in stanene. *Physical Review B*, 94, 4, 2016.
39. Matthes, L., Pulci, O. & Bechstedt, F., Massive Dirac quasiparticles in the optical absorbance of graphene, silicene, germanene, and tinene. *Journal of Physics Condensed Matter An Institute of Physics Journal*, 25, 39, 2013.
40. Lu, S. B. et al., Broadband nonlinear optical response in multi-layer black phosphorus: an emerging infrared and mid-infrared optical material. *Optics Express*, 23, 9, 2015.
41. Guo, Z., Han, Z., Lu, S., Wang, Z. & Chu, P. K., From black phosphorus to phosphorene: basic solvent exfoliation, evolution of raman scattering, and applications to ultrafast photonics. *Advanced Functional Materials*, 25, 45, 2016.
42. Wang, Y. et al., Electrical contacts in monolayer arsenene devices. *ACS Applied Materials & Interfaces*, 9, 34, 2017.
43. Zhang, S. et al., Semiconducting group 15 monolayers: a broad range of band gaps and high carrier mobilities. *Angewandte Chemie - International Edition*, 128, 5, 2016.
44. Gong, Fei et al., Bimetallic oxide mnmoo$_x$ nanorods for in vivo photoacoustic imaging of GSH and tumor-specific photothermal therapy. *Nano Letters*, 18, 9, 2018.
45. Song, Y. et al., Few-layer antimonene decorated microfiber: ultra-short pulse generation and all-optical thresholding with enhanced long term stability. *2D Materials*, 4, 4, 2017.
46. Cai, X., Han, X., Zhao, C., Niu, C. & Jia, Y., Tellurene: an elemental 2D monolayer material beyond its bulk phases without van der Waals layered structures. *Journal of Semiconductors*, 41, 8, 2020.
47. Zhu, Z. et al., Multivalency-driven formation of Te-based monolayer materials: a combined first-principles and experimental study. *Physical Review Letters*, 119, 10, 2017.
48. Boettger, J. C. & Trickey, S. B., First-principles calculation of the spin-orbit splitting in graphene. *Physical Review B*, 75, 19, 2007.
49. Wang, X. Q., Li, H. D. & Wang, J. T., Induced ferromagnetism in one-side semihydrogenated silicene and germanene. *Physical Chemistry Chemical Physics*, 14, 9, 2012.
50. Voon, L., Physical properties of silicene[M]//Silicene. Springer, Cham, 2016: 3–33.
51. Liu, L., Ji, Y. & Liu, L., First principles calculation of electronic, phonon and thermal properties of hydrogenated germanene. *Bulletin of Materials Science*, 42, 4, 2019.
52. Liu, Cheng-Cheng et al., Quantum spin hall effect in silicene and two-dimensional germanium. *Physical Review Letters*, 107, 7, 2011.
53. Lu, P. et al., Quasiparticle and optical properties of strained stanene and stanane. *Scientific Reports*, 7, 1, 2017.
54. Li, Y., Zhang, J., Zhao, B., Xue, Y. & Yang, Z., Origin of topologically trivial states and topological phase transitions in low-buckled plumbene. arXiv preprint arXiv:1810.02934, 2018.

2 Basic Structures, Properties, and Applications of Group IIIA

Kai Wang and Jun Li

CONTENTS

2.1 INTRODUCTION

Recently, Xenes have received increasing attention attributed of their superior characteristics, especially group IIIA. This is because an emerging two-dimensional (2D) material that Xenes have has been theoretically predicted to be used in electronics, sensing, biomedicine, chemistry, capacitors, batteries, etc. The fascinating properties that they have exhibited, such as electronic properties, mechanical flexibility, superconductivity, as well as in-plane anisotropy have also confirmed that they will become the mainstream of the next generation of materials science. At present, black phosphorus is the most widely investigated material among them and has been abroadly used in optoelectronic devices, and in catalytic and other fields.[1-4] However, to achieve the stability of black phosphorus at room temperature is still full of challenges. Even so, black phosphorus has

DOI: 10.1201/9781003207122-2

obtained extensive success, especially in the fields of energy and health care.[5, 6] In addition, other 2D Xenes materials have exhibited that they are different from the previously reported materials' similar transition metal dichalcogenides (TMDs). The electronic band structure of these have been demonstrated to be used in many ways to explore various applications. Typically, 2D Xenes have shown great advantages in the field of electrochemical energy due to their special folding structure and ultra-high surface-to-volume ratio and the exposed surface lone pair electrons of Xenes, making them easy to be chemically functionalized as well as large layer spacing and excellent flexibility.

Recently, the group IIIA elements, as a considerable potential 2D material (borophene, aluminene, gallenene, indiene) in energy, have received increasing attention due to their versatile properties. Recently, borophene, as the extensively studied material in group IIIA, has shown a variety of phase structures and potential applications in multiple fields. For instance, borophene, which is inert to oxidation, exhibits excellent mechanical, chemical, electrical, and thermal properties.[7-9] Meanwhile, borophene exhibits a vast potential in biosensing application for its large surface area and superb absorbability of gas molecules. To reveal the mechanism, Huang et al. analyzed the transportation characteristics of borophene after gas adsorption[10] by the means of NEGF-based methods. In addition, with the stable conductive properties, light weight, high absorbance, and the good photothermal responsiveness, borophene has illustrated potential usages as battery anodes, hydrogen storage, bioimaging, and in cancer therapy.[11-13] Especially the buckling structure has been proven to pave the way for the birth of topological insulators. For these monoelements, the method of preparation is very difficult because it is hard to maintain stability in them.[14] Gozar et al. have synthesized the borophene on Ag (111) and Cu (111). However, the obtained borophene nanosheets are too small to fabricate a device. Subsequently, Wu et al. achieved a major breakthrough in the synthesis of borophene with a long-awaited planar hexagonal honeycomb structure. They synthesize a honeycomb structure of borophene on an aluminum substrate. Hereafter, Teo et al. demonstrated the control and high quality of borophene successfully prepared by liquid-phase exfoliation. Notably, the synthesis of borophene is a breakthrough for the group IIIA materials. In addition to borophene, gallenene has also been proven to be successfully prepared by solid melt exfoliation. In terms of aluminene and indiene, they are now at more of a theoretical stage of inquiry. Although there has been some research on their preparation, this is still only at the initial stage. This means that there are lots of challenges to be overcome in the future.

Herein, we mainly introduce the current research on the structure, properties, and application of the group IIIA elements including borophene, aluminene, gallenene, and indiene, discussing their basic energy band structure, in-plane anisotropy, and their influencing factors. At present, the most urgent problem to be solved is the cost of preparation and the stability of the material. This will still be a big challenge in the near future. In addition, we summarize their effects in biology and energy, as well as the latest research progress in the fields of medicine, energy storage, electronics, etc.

2.2 BOROPHENE

2.2.1 BASIC STRUCTURES AND PROPERTIES

Structures: boron, which is the main element that makes up borophene, with unique bonding behavior is the only non-metallic element in group IIIA. Recently, borophene has been demonstrated to have little sensitive to oxidation due to the difference of valence orbital. Compared to carbon, boron, with an electron-deficient nature, forms a maximum of 16 volumes of allotropes, so the borophene cannot form a graphene-like honeycomb (Figure 2.1(a), (b)). Currently, borophene, as an elemental metallic Dirac material, has been confirmed to have at least two configurations, one of which is buckled triangle (δ_6) and the other is flat configuration (β_{12}), in which the network configuration of buckled triangle is proven to be more stable.[15] For flat configuration (β_{12}), which is unstable due to numbers of electrons occupying the anti-bonding orbital, it can maintain a stable state by the introduction of vacancies.

Lately, based on broad theoretical predictions, various forms of single-layer borophene structures have been proposed and fabricated. The statistics suggest

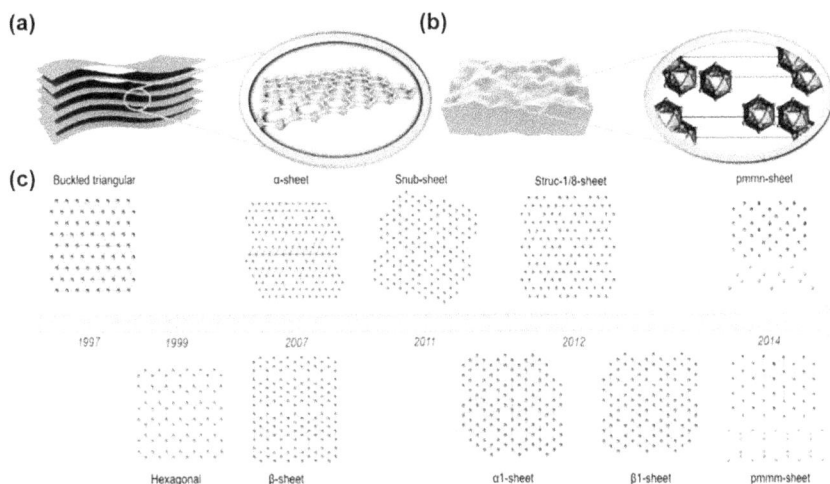

FIGURE 2.1 (a) Structure of bulk graphite. (b) Structure of bulk boron. (c) Development history of borophene in theoretical models. Buckled triangular structure of borophene was first predicted in 1997; then a novel boron single layer structure, α-sheet, consisting of triangular lattice and hexagonal honeycomb holes, was obtained through calculation. Boron atoms of α-sheet are in the same plane, the energy of which are lower than the buckled triangular structure that was considered to be the most stable in previous studies. Based on this, during 2007–2014, a series of classical borophene structures such as β-sheet, snub-sheet, α_1-sheet, struc-1/8-sheet, β_1-sheet, Pmmn-sheet, and Pmmm-sheet occurred through the different arrangement of hexagonal and triangular lattices. The energy of these structures is very low and almost equal. Reproduced with permission.[95] Copyright 2021, Wiley-VCH.

that growth environment and the choice of substrate are significant for the exploration of borophene phase.[8, 15–23] The hexagonal atomic structure[24] is a representative structure of monolayer borophene. The in-plane and out-of-plane thermal conductivities of monolayer and multilayer borophene were studied based on this structure.[25] According to first-principles calculations, the hexagonal lattice constant and buckling distance of the borophene structure are: a = 2.866 Å, b = 1.614 Å, and h = 0.911 Å. Meanwhile, the research illustrates that the in-plane thermal conductivity of single-layer borophene is easily affected by temperature, size, and mechanical strain. Figure 2.1(c) reveals the discovered structure of borophene according to year.

Properties: as the lightest anisotropic Dirac material, the emergence of borophene is phenomenal, which is mainly attributed to its unique electronic and mechanical characteristics. Certainly the other unique characteristics are also necessary factors that need to be considered.

Electronic characteristics: with the electron-deficient nature of B element, the electron band structure of multilayer borophene is intersected by more bands than the monolayer because of band splitting. Therefore, borophene still retains obvious metallic properties as the number of layers increases. Although different from the stable structure of graphene, borophene also has many unique electrical properties, such as superconductivity, Dirac-Fermi effect, metallicity, semiconductivity, and superconductivity. Experiments have confirmed that both β_{12} and χ_3 borophene are Dirac hosts. The difference is that borophene is completely hydrogenated, showing a twisted Dirac cone which has a Fermi velocity four times higher than that of graphene. Evgeni et al. theoretically predicted that the various structural polymorphs of borophene are metallic. Moreover, the small atomic mass of borophene suggests strong electron-phonon coupling, which in turn can enable superconducting behavior.[9] However, electron doping and tensile strain can lead to inhibition of borophene superconductivity, making it tough to implement the critical temperature of borophene in experiments. Due to the non-zero band gap, several phases of borophene have been experimentally confirmed by scientists to be semiconducting. Kou et al. found that we can change the band gap value of the boron surface by changing the applied stress and transform the conductive boron benzene from metal to semiconductor.[26]

Mechanical properties: borophene is composed of multi-center covalent bonds instead of the classic double-center covalent bonds that exist in other ordinary 2D materials. It has excellent mechanical properties. exploring the ideal tensile strength and critical strain by applying tensile strain to uniaxial (armchair or zigzag) or biaxial directions. A method similar to the study of 2D black scales confirmed that after each uniaxial strain is applied, the lattice constants along the transverse direction and boron atoms are completely relaxed. For biaxial strain, applying equal biaxial tension, the boron atoms in the unit cell are completely relaxed. Apart from these potentially interesting properties above mentioned, borophene also exhibits in-plane anisotropy in different directions and the optical property. It has been predicted to have a high degree of transparency, contributed to its extremely thin thickness. This may make it a strong candidate for photovoltaic materials and touch screens. Furthermore, the absorption and reflection of borophene are also anisotropic. It exhibits high reflectivity in the visible light region, and as the number of layers changes, the transmission area will also change accordingly.[25, 27–29] In fact, borophene is also predicted to be a primitive 2D mono-layer superconductor whose superconductivity can be significantly enhanced by straining or doping. It is confirmed by means of tensile strain and hole doping, where borophene grown on a substrate with larger lattice parameters or under light excitation can exhibit enhanced superconductivity with critical temperature above liquid hydrogen temperatures of 20.3 K.[9, 30]

Chemical property: 2D boron is inert to oxidation, while its outer edge and 3D bulk boron are different. Since the number of boron valence electrons is less than the number of valence orbitals, when a chemical substance is formed, boron atoms with electron deficient will be formed. Wu et al. confirmed the oxidation of boron mainly occurs at the edges, and the boron nanosheets themselves are stable. Therefore, compared with bulk boron, 2D borophene is more inert to oxidation.[8, 20, 31–33] Furthermore, by calculating the observed peak area of boron oxide, the study found that most of the boron atoms on the surface of borophene have anti-oxidation stability. Meanwhile, Guisinger et al. found that the oxidation state of borophene detected mainly occurs at the edge because of its active edge states.[34] In addition, due to the chemical activity at the edge of borobenzene, it could also be used to catalyze hydrogen evolution. This fascinating feature opens up a new way for the application of borophene.[7]

Thermal properties: borophene has excellent thermal properties, including thermal stability and thermal conductivity. In recent years, studies have shown that the thermal properties of borophene are closely related to the structure of borophene. Recent studies on borophene suggest that borophene exhibits special properties in heat transfer, and this property varies depending on the structure. Based on the borophene currently obtained on Ag (111) and Cu (111) substrates, once removed from the substrate, they may be unstable, especially the δ_6 phase. To eliminate the unstableness of borophene, producing fully hydrogenated borophene is a strategy we can choose.[35, 36]

Generally, when the conductor is in a temperature gradient, the phonon transport and scattering mechanism of the thermal transport process is complicated.[37] Through non-equilibrium Green's function (NEGF) and ab initio calculations, it is found that borophene exhibits superior lattice thermal conductivity in the ballistic region, surpassing graphene. This may be due to the light mass, short B-B bond, abnormal phonon transmission, and the strong structural anisotropy of borophene, resulting in higher phonon group velocity.[38–40] The phonon frequencies of the various borophene phases are close to each other because borophene is composed of the same boron atoms. Nevertheless, the difference in the number of boron atoms in different unit cells leads to different geometrically symmetric structures and different numbers of optical phonon branches, which form multiple heat transports.[37, 41]

2.2.2 SYNTHESIS AND APPLICATIONS

Borophene: actually, as a material synthesized, more research has been done on borophene's preparation methods and application compared with aluminene, gallenene, or indiene. Recently, the manufacturing of 2D borophene flakes on different metal substrates has achieved great breakthrough. For instance, Wang et al. prepared the boron nanosheets growing on the Cu (111) surface by a homemade two-part CVD boiler (Figure 2.2(a)). The 8-Pmmn-structured borophene film obtained in this method is large and continuous (Figure 2.2(e), (d)), and it was confirmed to be a single-layer film by the AFM (Figure 2.2(f)) and TEM (Figure 2.2(g)).

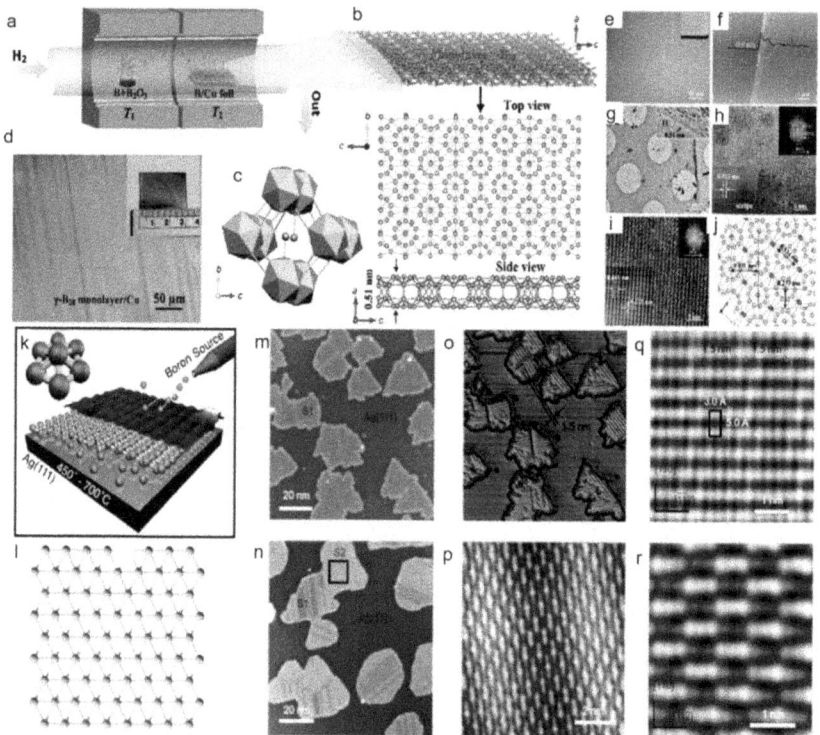

FIGURE 2.2 (a–d) Schematic diagram of the synthesis of borophene and the structure of borophene in a tube furnace. (e) Optical image. (f) AFM. (g–j) TEM. (a–j) Reproduced with permission.[96] Copyright 2015, *Nature Science*. (k) Schematics of synthesizing borophene on silver. (l) Structure of borophene predicted by computer. (k, l) Reproduced with permission.[97] Copyright 2015. (m–r) Closed-loop dI/dV images of borophene on the right and large-scale STM topography of borophene on the left, displaying low medium. (o) The striped-phase nanoribbons, striped-phase islands, and homogeneous-phase. (p) STM images about atomic-level structure of striped-phase. The rectangular lattice shown in the illustration has overlaid lattice vectors. (q) The arrow and rhombus denote honeycomb and rhombohedral Moiré patterns in striped phase. (r) Carpet-mode growth demonstrated by striped-phase island. The illustration indicates that atoms across Ag (111) step have continuity. (m–r) Reproduced with permission.[98] Copyright 2016, *Springer Nature*.

Immediately afterwards, Feng et al. obtained boron on Ag (111) substrate. They found that when the substrate temperature is lower than 500 K, only clusters or no structure can be formed. When the temperature reaches 570 K, boron with a perfectly ordered structure is obtained. Figure 2.2(k–l) exhibits the preparation diagram on the silver substrate and the calculated structure diagram. And Figure 2.2 (m–r) shows the two phases (S1 and S2) we observed, which are demonstrated corresponding to the theoretical predictions β_{12} and χ_3. Experiments indicated that the formation and stability of different borophene structures are greatly dependent on the growth substrate. At the same time, temperature plays a decisive role in the appearance of the borophene phase and the transition between the phases.[14, 42–45] This means that we can get the phase structure we need by changing the experimental conditions to meet the need of application. Not long ago, 2D borophene was successfully synthesized by Teo et al. through a liquid-phase exfoliation (Figure 2.2(s)). Subsequently, the sample was characterized by AFM (Figure 2.2(t)). The average thickness of borophene exfoliated in dimethylformamide is 1.8 nm and the area is 19,827 nm^2, while the thickness of borophene exfoliated in isopropanol is 4.7 nm and the area is 1791 nm^2; therefore, the size and thickness of borophene can be controlled by adjusting the solvent used in ultrasonic peeling.

Figure 2.3(e) shows the TEM image of already prepared B flakes by the other way of liquid-phase exfoliation (Figure 2.3(a)) by Ji et al. Since ultrasonic energy can break the interlayer valence bonds in a possible chain-by-chain manner, it is possible to produce a few layers of borophene. Figure 2.3(b) demonstrates the thickness of B flakes of about 3 nm.[46] Compared with nanosheets prepared on a metal substrate, the 2D boron sheets prepared by the liquid phase exfoliation method are larger. Furthermore, the synthesis of borophene polymorphs by hydrogenated borophene has been reported by Li et al. Similar to the high polymorphism in borophene, eight different borophene polymorphisms were observed. Significantly, borophene polycrystalline overcomes the disadvantages of immediate oxidation of borophene when exposed to ambient conditions.[34, 47]

In numerous previous reports, it was theoretically predicted that borophene can be used as a superior anode metal ion battery material, and its obvious application potential was illustrated in hydrogen storage, supercapacitors, sensors, as well as biomedicine owing to its excellent chemical, electronic, mechanical, and thermal properties.[22, 48–51] As the current production of borophene is full of challenges, its application in electronics has been largely restricted, but it has shown great advantages in biomedicine. Among these, the nanosized materials are favored in the field of tumor treatment.[52, 53] We report the latest application progress of borophene here.

Batteries: metallicity and low atomic weight make borophene promising as a candidate material for battery photoanode. Meanwhile, theoretical research has confirmed that when the hollow hexagonal vacancies in borophene are adsorbed on metals such as lithium and sodium, the whole system can remain stable or even enhanced.[13]

The boron atoms of buckled borophene can be divided into two groups according to their position height. The loose, layered, stacked structure can maintain a small volume expansion and lattice change after lithium and sodium plasma

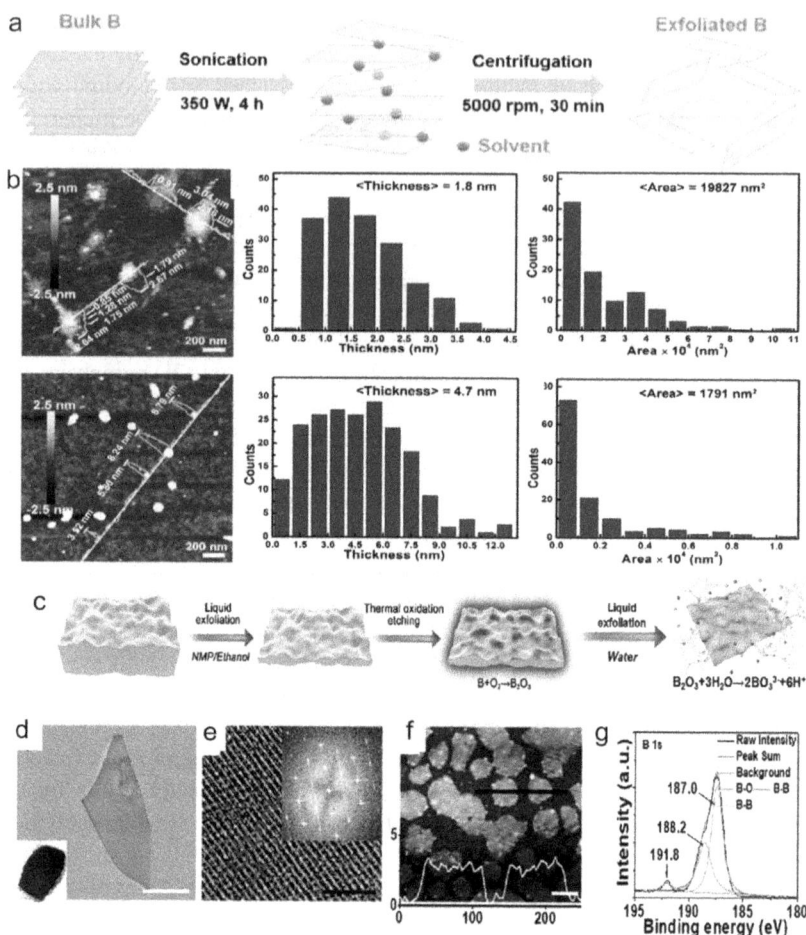

FIGURE 2.3 (a) Schematic of the sonication-assisted liquid-phase exfoliation synthesis of high-quality borophenes. (b–g) AFM characterization of borophene synthesized in DMF (b–d) and IPA (e–g) by sonication. (b, e) AFM topographic images and height profiles. (c, f, d, g) Thickness and size of prepared materials. h) Schematic of thermal oxidation etching and liquid exfoliation synthesis of borophene. (d–f) TEM, HRTEM, AFM images, and XPS spectra of borophene synthesized by liquid exfoliation in water. (a, b) Reproduced with permission.[99] Copyright 2018, American Chemical Society. (c–g) Reproduced with permission.[100] Copyright 2018, Wiley-VCH.

intercalation. Borophene can ensure good cycle stability and an almost complete structure when used as a battery photoanode. Not only that, the theoretical capacity value of a sodium-ion battery containing boron benzene reaches 1218 mAh g^{-1} $Na_{0.5}$ B, which is three times that of graphite anode. In the case of lithium, it is even better. Thus, borophene can greatly increase the metal embedding effect and increase the battery capacity density. Furthermore, Xu et al. found that the average open circuit voltage of sodium-ion batteries using borophene as the cathode material during the charging process is as low as 0.53 volts during charging, and the bond between borophene and Na is strong, as shown Figure 2.4(a).[54] Both show that borophene is uniformly covered by metal atoms, which can effectively inhibit the formation of dendrites while maintaining high energy density.

Although borophene has been shown to be a very promising candidate material as a battery anode, at the moment it is still only at the theoretical stage and will still need a lot of experiments to confirm. In addition, it also shows great potential in biomedicine.

Drug delivery: reducing the side effects of chemotherapy is very critical, and the local release of chemotherapy drugs is precisely the most influential in this regard. Therefore, it is of great significance to build such a special platform. Borophene has an ultra-high specific surface area, which provides space for drug loading and an anchor point for functional group modification, making it suitable as a candidate material for anti-cancer therapy. In a system that uses borophene as the main body to deliver drugs, doxorubicin is co-cultured with borophene (loading rate 114%), and the system is coated with polyethylene glycol-NH_2 to achieve long-term circulation of chemotherapeutics, as shown Figure 2.4(b). As a qualified drug carrier, borophene exhibits excellent pH and photothermal responses, which support borophene to achieve targeted drug release at the tumor. The tumor microenvironment is lower than the pH of normal tissues.[51]

Biosensors: as a device with a high degree of specificity and sensitivity, biosensors are used to detect basic biological substances such as hormones, proteins, and glucose. 2D materials, with excellent electrochemical performance, multiple active sites, and high electron mobility, are considered as candidates for biosensors.[55–57]

Borophene exhibits excellent adsorption to gas molecules due to its noticeably large specific surface area. Regardless of whether the gas is toxic or not, biosensors based on borophene are immune. The electronic band gap value of borophene will be reduced accordingly when gas is adsorbed, which is conducive to the transfer of a large number of electrons and improves the conductivity. As a well-known irritant and highly toxic carcinogen, formaldehyde will cause a series of adverse reactions after contact with it. However, it is widely favored in the chemical and pharmaceutical industries because of its good high-thermal stability. Ansari et al. found that the presence of formaldehyde can increase the conductivity of B_{36}, generate electrical signals, and allow analysis of borophene's hexagonal hollow structure through calculations. In terms of application, this confirms the outstanding performance of borophene in the detection of certain toxic gases.[10, 58–60]

In addition, Huang et al. used the basic method to adsorb nitric oxide (NO) and carbon monoxide (CO) on the transport characteristics of boron benzene, which proved that the extremely high adsorption energy on the boron benzene surface will lead to a large amount of charge transfer. CO and NO have good prospects in anti-inflammatory, anti-bacterial, and anti-tumor aspects. Borophene biosensors make their transmission controllable and safe. Although the borophene biosensor

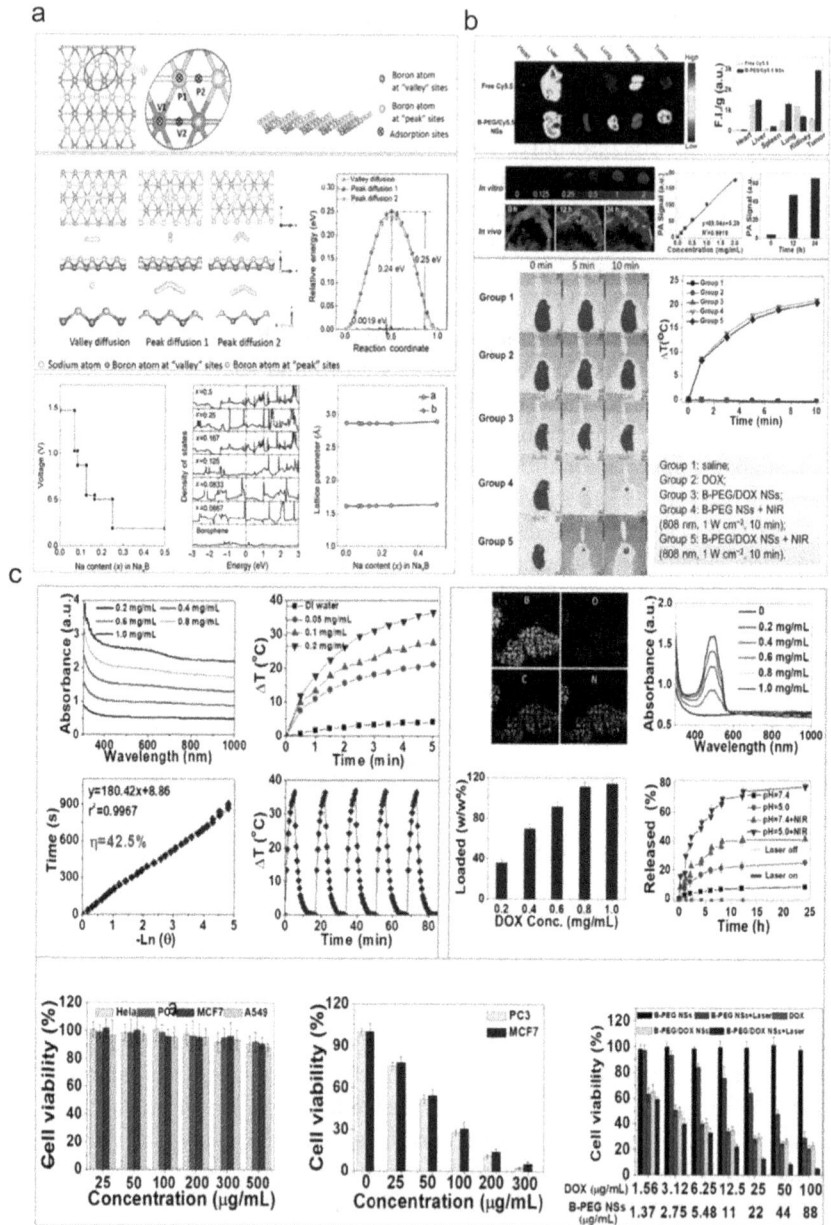

FIGURE 2.4 (a) Top view, side view, and the chosen adsorption sites of Pmmn borophene. Three sodium diffusion routes on borophene. The energy curves of diffusion routes. Voltage change during the sodiation process. Borophene's DOS and lattice parameters under various sodium concentrations. (a) Reproduced with permission.[54] Copyright 2016, Elsevier. (b) Multimodal imaging of B-based NSs. Detecting major organs and tumors with fluorescence imaging. Semiquantitative biodistribution of major organs and

has been proven to be an outstanding gas detection device, its stability will be greatly reduced when the gas is adsorbed, which is a problem that needs to be solved urgently. Therefore, for stable, sustainable, and safe applications in the medical field, more experimental evidence would be indispensable.[61]

Hydrogen storage: based on limited fuels and environmental protection issues, it is quite urgent to develop new candidates for hydrogen storage materials. Studies have found that for pure borophene, its maximum adsorption energy for H_2 is less than 0.01 eV, so it is not suitable for use as a hydrogen storage material. Fortunately, with the assistance of alkali metals or transition metals, the chemical activity of borophene can be enhanced.[62] Tang et al.[63] found that borophene modified by calcium is a perfect candidate for hydrogen storage. Furthermore, borophene does not need to produce defects to adsorb metals, because its HHS can be used as metal adsorption sites.[12, 64, 65]

It has been proven that light metal elements can greatly improve hydrogen storage capacity. For instance, Li-modified β_{12} borophene can reach 10.85% by weight of H_2 adsorption capacity.[66] The lithium-modified β_{12} structure borophene can reach 10.85% by weight of H_2 adsorption capacity, while a-sheet modified Li can adsorb up to 10.75% by weight of H_2, so each lithium atom can adsorb up to $3H_2$ molecules.[67, 68]

Bioimaging: bioimaging, which includes photothermal imaging, photogenerated imaging, and fluorescence imaging, is mainly used in medicine to observe the position and size of tumors and is an important technology for clinical diagnosis and treatment of tumors.[68–71] Each imaging mode has its pros and cons. Therefore, the multi-mode imaging system is relatively more comprehensive, which can greatly reduce the chance of misdiagnosis.

In past reports, borophene has been used in a tumor multi-mode imaging platform constructed by Ji et al.,[51] which can perform multiple imaging modes such as fluorescence imaging and photothermal imaging at the same time. In Figure 2.4(c), the fluorescence imaging picture of a mouse that has been injected with Cy5.5-labeled borophene tumors modified with polyethylene glycol clearly distinguishes malignant tumors from normal tissues.

FIGURE 2.4 (CONTINUED)

tumors. PA images of B-PEG NSs with different concentration (0, 0.125, 0.25, 0.5, 0.1, and 0.2 mg mL^{-1}) in vivo. Tumor PA images. Linear relationship between PA values and concentration of B-PEG NSs. PA values' quantitative analysis. Photothermal images and temperature profile of mice. (b) Reproduced with permission.[100] Copyright 2018, Wiley-VCH. UV spectrum of B-PEG NSs. Temperature-change profile of B-PEG NSs under 808 nm laser. Linear relationship between-lnθ and time. Heating profile of B-PEG NSs after five cycles. Scanning transmission electron microscopy-EDS images of B-PEG/DOX NSs. (c) UV spectrum of borophene loading with DOX. Borophene's ability to load chemotherapy drugs. Release curves of DOX under different pHs. Cell experiments. Safety evaluation of borophene in different cells (without laser). Toxicity evaluation of borophene in different cells (808 nm laser). Relative viability of tumor cells under different treatments. (c) Reproduced with permission.[100] Copyright 2018, Wiley-VCH.

2.3 ALUMINENE

2.3.1 STRUCTURE AND PROPERTIES

Structure: for the remaining three metal elements from group IIIA, honeycomb structure-like are ubiquitous among them. Aluminene exhibits the greatly difference compared with how other honeycomb materials behave as semiconductors or semi-metals, because it is a metal partially filled with σ and π bands. Currently, the buckling phase and 8-Pmmn aluminene have been proven to be dynamic and thermodynamically stable by Yuan et al.[72, 73] Besides that, the research shows that buckling aluminene is mainly through flexion interlayer and a strong metallic bond layer covalently bonded, while 8-Pmmn aluminene mainly through an aluminum-a covalent bond between aluminene atoms bonded. In Figure 2.5(a), two configurations of aluminene are shown, in which each honeycomb unit in the buckling phase contains two equivalent aluminum atoms. For the symmetrical 8-Pmmn aluminene, these are named inner atoms and ridge atoms. In order to gain better insight into the chemical bonding of aluminene, we plotted the charge density difference between aluminene and its atomic constituents.

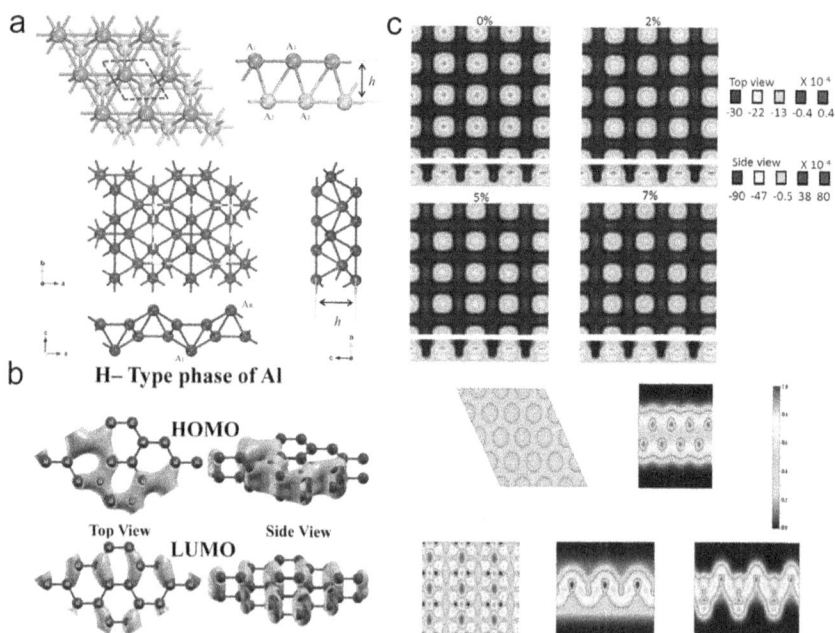

FIGURE 2.5 (a) The optimized geometric structures of buckled and 8-Pmmn aluminene. top and side view of buckled aluminene. Top view and side view of 8-Pmmn aluminene. Dashed rectangles delimit the unit cell. (b) Visualization of HOMO and LUMO structure for bilayers of a. (b) Reproduced with permission.[81] Copyright 2018, American Chemical Society. (c) Charge density difference between aluminene and its atomic constituents. (c) Reproduced with permission.[80] Copyright 2018, Elsevier.

Properties: in the current research, the four configurations are predicted based on the density functional theory: plane, buckling, fold, and triangular geometric configuration plane. The study found that among these four configurations, only the planar honeycomb monolayer structure made of aluminum is stable, just like graphene.[74]

Stability: inspired by borophene, aluminene may also be an unknown Dirac material; in view of the few reports of aluminene at present, it is quite necessary to study the stability of aluminene for experimental fabrication and large-scale production. Yuan et al. calculated the phonon dispersion of its geometric configuration, and showed the stability of the two configurations (buckled and 8-Pmmn phases) that have been reported. In addition, no spin polarization is observed in any of the structures, which means that both configurations are non-magnetic. Furthermore, calculating the formation energy of isolated aluminene nanosheets, the phonon density of states, and phonon dispersion calculations confirm its structural stability due to the stability of 2D aluminene-based materials that are crucial for experimental manufacturing.[74, 75]

Bonding characteristics: in the 8-Pmmn aluminene structure, no obvious Dirac cones are observed at the Fermi level, which is the opposite of boron benzene. Therefore, the study of its bonding characteristics is very meaningful. As shown Figure 2.5(c), calculating the electronic regionalization function (ELF) is a universal method to understand the bonding characteristics. The ELF value (~0.5) indicates that the buckled aluminum atoms are typically metal bonds, while the aluminum atoms in 8-Pmmn aluminene are stronger covalent bonds due to electrons located near aluminum atoms along the (100) and (010) plane.[74]

Electronic properties: as a single-element material similar to graphene, aluminene can also form a honeycomb structure. The honeycomb materials reported so far are either semiconductors or semi-metals. Therefore, the study of aluminene electronic properties is quite constructive. Figure 2.5(b) explores the electronic properties of aluminene, based on highest occupied molecular orbital (HOMO), lowest unoccupied molecular orbital (LUMO), and density of states (DOS) spectrum.[76] From the results of quantum chemistry calculations, the isolated aluminene nanosheets exhibit metalloid properties according to their energy gap value of 0.302 eV. In addition, it was also found that the adsorption of gas molecules can widen the band gap value of aluminene nanosheets.[74]

Thermodynamic property: thermal vibration plays an important role in the heat transfer of materials. For this emerging material, aluminene, the exploration of its thermodynamic properties will have a major impact on future applications in the field of electronic devices. Regarding thermal vibration, Debye temperature is a measure of the temperature at which all modes start to be excited, which will further affect heat transfer in aluminene. By calculating the tobah temperature, it is found that the tobray temperature of buckling aluminene is higher than that of single-layer MoS_2[77] and arsenene,[78] but lower than that of black phosphorus, borophene, and graphene.[18, 79] For 8-Pmmn aluminene, the debye temperature is much lower than buckling aluminene.

Superconductivity: aluminene has been predicted to have the possibility of being produced in experiments. After confirming its stability, Yeoh et al. discovered its superconductivity. A comprehensive analysis of the electron-phonon coupling in the structure of aluminene revealed that the pristine form of aluminene can superconduct at a superconducting temperature of 6.5 K. Under biaxial tensile strain, the superconducting critical temperature is further increased to 11.9 K, which is mainly attributed to the strong electron-phonon coupling produced by the density of states of the Fermi level as the tensile strain increases.[80]

2.3.2 SYNTHESIS AND APPLICATION

Currently, the study of aluminene is still at the stage of theoretical research. In a 2018 report, the homogenized junction of aluminene and antimony has been shown to be a promising material for charge storage as nanocapacitors.[81] The transformation of electrons plays a major role in understanding the behavior of capacitive materials. Therefore, the dipole moment of the smallest electron transfer between layers is studied. In addition, aluminene has also been predicted and theoretically proven to be a potential hydrogen storage material because of its high specific surface area.[82] By introducing low-concentration calcium, potassium, and magnesium ions into planar aluminene, density functional theory is used to determine its effect on hydrogen adsorption. Remarkably, the bonding of Mg, K, and Ca does not change the non-magnetic metal properties of aluminene. The interaction mechanism between impurity and aluminene surface is ionic.

2.4 GALLENENE AND INDIENE

2.4.1 STRUCTURE

It is well known that gallium, as a liquid metal, has six different phases and has a different degree of metallicity at room temperature. As shown Figure 2.6(a), the optimized double-layer gallium and triple-layer gallenene can minimize the recombination in the structure. Different from the reported formation of borophene, gallenene is formed by cutting covalently bonded dimers. And it could make the metallicity in the plane unchanged. In addition, gallenene can be covalently bonded to the substrate material because it has dangling covalent bonds.[84]

As for indiene, three structures have been reported so far: planar, puckered, and buckled. Singh et al. calculated the total ground state energy of the three structures through first principles, and predicted that the three structures are all stable. Additionally, the allotrope of the planar structure can form a graphene-like honeycomb structure, while the other two structures cannot. Among them, for the puckered structure, the relaxed lattice constants are a = 4.25 Å, b = 5.68 Å, the planar structure is 4.96 Å, and the buckling structure is 4.24 Å. The optimized geometry is placed in Figure 2.6(d). By comparing with similar elemental monolayer structures, it shows that they follow the same trend as bulk materials.

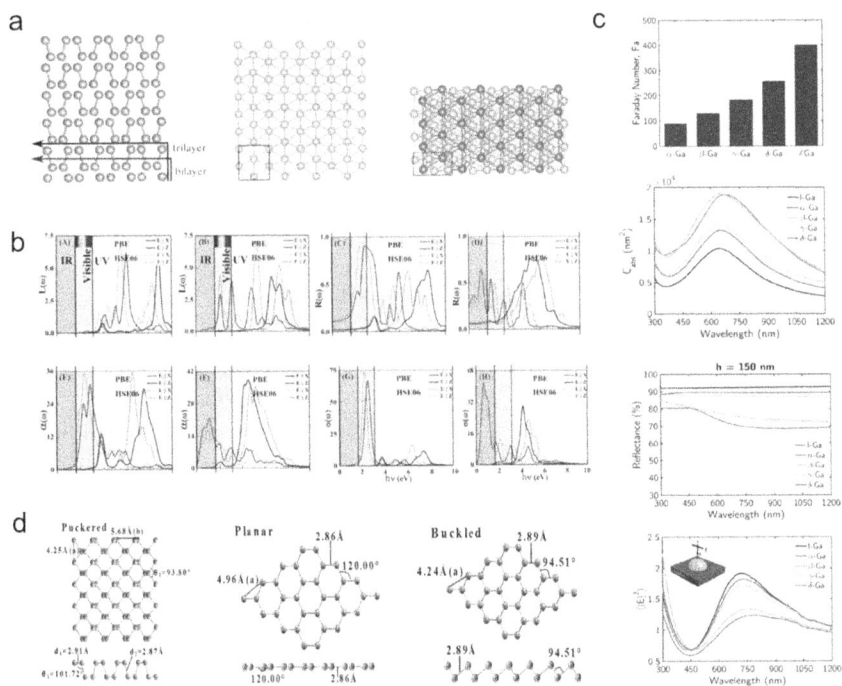

FIGURE 2.6 (a) Bulk α-gallium, optimized gallium bilayer structure and optimized gallium trilayer structure. (a) Reproduced with permission.[101] Copyright 2019, Royal Society of Chemistry. (b) Evaluation of plasmonic performance of thin films and nanostructures. Faraday number (Fa) of the different Ga phases. Reflectance spectra at normal incidence of a 150-nm-thick layer of Ga in the different phases on a sapphire substrate (α-Al_2O_3, n = 1.78). Absorption cross section (Cabs) and near-field enhancement ($|E|^2$) averaged over the surface of different Ga phase hemispheres of radius R=60nm deposited on a sapphire substrate (α-Al_2O_3, n = 1.78) illuminated under normal incidence. (c) The frequency dependent energy loss spectrum of buckled and planar indiene reflectivity of buckled and planar indiene, absorption spectra of planar and buckled indiene optical, and conductivity of buckled and planar indiene. (b, c) Reproduced with permission.[85] Copyright 2016, Royal Society of Chemistry. (d) Optimized geometries of puckered, planar, and buckled indiene. The lines indicate the equilibrium lattice constants. (d) Reproduced with permission.[102] Copyright 2020, Royal Society of Chemistry.

Therefore, monolayer indiene is larger than flat aluminene attributed to the fact that the lattice constant of bulk indium is larger than that of aluminum.[85]

2.4.2 Generality and Individuality

Stability: similar to aluminene, the newly discovered gallenene is also worthy of attention. Lambie et al. reported the thermal stability of 2D Ga. The ab initio molecular dynamics (AIMD) simulation found that the four-layer system melted at 457 K, the five-layer system was at 350 K, and the six-layer system was at

433 K. It concluded that four to six layers of 2D gallium are thermally stable.[86] Furthermore, compared with odd-numbered layers, even-numbered layers have higher stability. Currently, with the reported indiene structures, verifying their stability is of great significance for large-scale preparation in experiments and application in various files. Singh et al. demonstrated the structural stability of 2D indiene. For puckered, planes, and buckling allotropes, the calculated total energy is −53.73, −53.66, and −53.69 eV/atom, respectively. In all cases the energy of indiene is almost equal. Therefore, the three structures are stable in terms of total energy. Furthermore, the stability of the structure is confirmed by calculating the full phonon dispersion curve,[85, 87] which is shown in Figure 2.5(b). The puckered structure shows the imaginary frequencies of the acoustic and some optical phonon modes in the entire Brillouin zone. Therefore, it is not an allotrope with stable lattice dynamics of indium, so it is difficult to obtain.

Anisotropy: the electronic structure of gallenene film exhibits a high degree of anisotropy. Unlike graphene, gallenene has one less valence electron than graphene, which means that an extra electron is needed to fill it. This phenomenon is similar to borophene. The sp^2 hybrid orbital in gallenene is closer to the Fermi level, which leads to anisotropy in the low energy band. Not only that, gallenene has also been demonstrated to have a wide range of optical properties, ranging from ultraviolet to terahertz. Furthermore, studies found that the thermal conductivity of gallenene is less than 1 W/mK, which is lower than other similar 2D materials such as graphene (2000 to 5000 W/mK at 300 K), silylene (5 to 50 W/mK at 300 K), etc.[86, 88–91]

Optoelectronic properties: based on the interesting properties exhibited by two-dimensional materials, 2D indiene has also received extensive attention. Singh et al. reported its optoelectronic properties to prove its application potential in the field of electronic devices. The electronic structure calculated by density functional theory (DFT) shows that planar indiene is metallic, while buckled indiene is semiconducting, with a band gap value of 1.6 eV. Not only that, the planar indiene structure has achieved a reflectivity of up to 90% in the visible light region, which is better than borophene. Meanwhile, in the ultraviolet to infrared region, buckled indiene also exhibits a light absorption coefficient that is significantly better than that of bulk indium. Besides that, after increasing the electric field, it is found that the light absorption coefficient of the monolayer indiene has been significantly improved.[85, 87]

2.4.3 SYNTHESIS AND APPLICATION

Recently, 2D gallium, which was named gallenene, has been successfully prepared on silicon using solid-melt exfoliation by Kochat[86] et al. Due to the large thermal vibration, the increase in temperature will cause the strength of the metal to decrease. When the strength is as low as close to the melting point of the solid substrate, the temperature difference between the liquid and the substrate will cause heterogeneous nucleation at the substrate-liquid metal interface (Figure 2.7(a–f)). The calculation of the interaction energy between gallium atoms shows that there

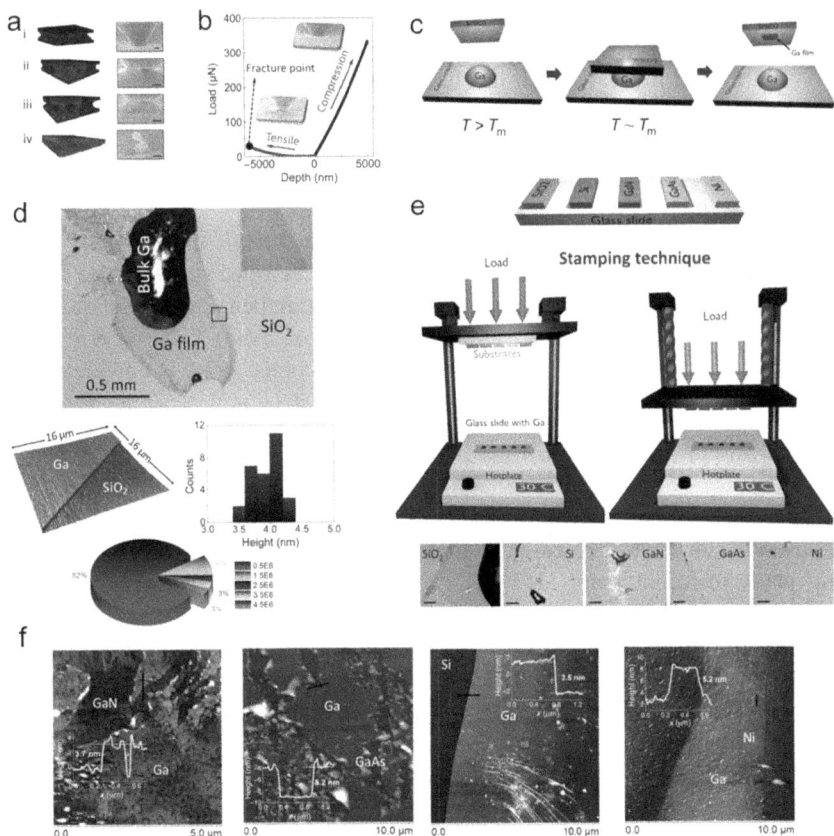

FIGURE 2.7 Solid-melt exfoliation of gallenene. (a) Snapshots of real-time imaging of gallenene exfoliation using a flat punch indenter inside SEM. The corresponding atomistic schematic of the fracturing is also shown. Scale bar, 1 μm. (b) Load versus displacement curve obtained from the in situ compression and tension test on molten Ga inside SEM. The inset reveals SEM images during tensile and compression loading of the indenter. (c) Schematic of the proposed solid-melt exfoliation technique of gallenene onto Si/SiO$_2$ wafers. (d) Optical image of Ga sheet on SiO$_2$ wafer showing regions with uniform ultra-thin layers. The AFM measurements on the films reveal the thickness of this film to be ~4 nm, as shown from the histogram of step height at the edge along different line scans. The pie chart shows the distribution of the percentage of the area of the exfoliated Ga films obtained from among 30 different flakes. (e) Schematic of the stamping technique developed for exfoliation on multiple substrates for a fixed load. Optical micrographs of gallenene flakes obtained on various types of substrates are shown at the bottom. Scale bar, 100 μm. (f) AFM images showing exfoliation of gallenene on various substrates like GaN, GaAs, and Si. (a–f) Reproduced with permission.[103] Copyright 2018, American Association for the Advancement of Science.

is a strong interaction between gallium and the matrix, which is different from van der Waals. Since the light absorption of liquid gallium involves the range from ultraviolet to terahertz, it has been used in photonic devices. Researchers introduced two gallenene structures with different atomic arrangements, and the results show that these structures are metallic.[92] Based on the optical and low thermal conductivity characteristics of gallenene, it is considered a potential thermal barrier in on-chip electrical connectors or devices.

Indiene is a 2D material in the same family as borophene, and current research is still stuck in the theoretical side. At present, the electronic structure and optical properties of indiene are mainly reported, and planar indiene shows a unique maximum value in the visible spectrum, as shown Figure 2.6(d). Compared with the buckling structure, indiene is an obvious anti-reflection coating in optoelectronic devices and exceptional candidate materials for electroluminescent devices.[93, 94]

2.5 SUMMARY

Group IIIA of two-dimensional Xenes monoelement materials has greatly aroused the interest of researchers due to the extensive research of borophene, especially in recent years. However, most of the current studies are still at the theoretical stage, and group IIIA materials have been predicted theoretically as superior candidate materials in the fields of batteries, sensing, optoelectronics, and biomedicine.

In addition, researchers have been intrigued by the outstanding performance of black phosphorous in the field of optoelectronic devices and catalysis as reported by previous studies and the theoretical prediction of silicene and antimonene as promising energy storage devices. For the group IIIA single element with prominent theoretical performance, they have shown a great advantage in energy storage. Overall, based on borophene, the anisotropy, electronic properties, and optical properties currently exhibited by aluminene materials have involved multiple fields, and there has been substantial progress in their research. But the challenge still exists. Therefore, more experimental support is necessary. More importantly, the primary matters now are the large-scale and high-quality preparation and stability under environmental conditions.

REFERENCE

1. Zhu, H. et al., Doping behaviors of adatoms adsorbed on phosphorene. *Physica Status Solidi B-Basic Solid State Physics*, 253, 6, 2016.
2. Huang, M. et al., Broadband black-phosphorus photodetectors with high responsivity. *Advanced Materials*, 28, 18, 2016.
3. Abellán, G. et al., Noncovalent functionalization and charge transfer in antimonene. *Angewandte Chemie - International Edition*, 56, 46, 2017.
4. Nakhanivej, P. et al., Revealing molecular-level surface redox sites of controllably oxidized black phosphorus nanosheets. *Nature Materials*, 18, 2, 2019.
5. Tao, W. et al., Black phosphorus: black phosphorus nanosheets as a robust delivery platform for cancer theranostics (Adv. Mater. 1/2017). *Advanced Materials*, 29, 1, 2017.

6. Srivastava, P. K. et al., Resonant tunnelling diodes based on twisted black phosphorus homostructures. *Nature Electronics*, 4, 4, 2021.
7. Zhang, Z., Penev, E. S. & Yakobson, B. I., Polyphony in B flat. *Nature Chemistry*, 8, 6, 2016.
8. Feng, B. et al., Experimental realization of two-dimensional boron sheets. *Nature Chemistry*, 8, 6, 2016.
9. Penev, E. S., Kutana, A. & Yakobson, B. I., Can two-dimensional boron superconduct? *Nano Letters*, 16, 4, 2016.
10. Huang, C.-S., Murat, A., Babar, V., Montes, E. & Schwingenschlögl, U., Adsorption of the gas molecules NH_3, NO, NO_2, and CO on borophene. *The Journal of Physical Chemistry C*, 122, 26, 2018.
11. Jang, B., Park, J.-Y., Tung, C.-H., Kim, I.-H. & Choi, Y., Gold nanorod–photosensitizer complex for near-infrared fluorescence imaging and photodynamic/photothermal therapy in vivo. *ACS Nano*, 5, 2, 2011.
12. Chen, X., Wang, L., Zhang, W., Zhang, J. & Yuan, Y., Ca-decorated borophene as potential candidates for hydrogen storage: a first-principle study. *International Journal of Hydrogen Energy*, 42, 31, 2017.
13. Kadioglu, Y., Ersan, F., Gökoğlu, G., Aktürk, O. Ü. & Aktürk, E., Adsorption of alkali and alkaline-earth metal atoms on stanene: a first-principles study. *Materials Chemistry and Physics*, 180, 2016.
14. Wu, R. et al., Large-area single-crystal sheets of borophene on Cu(111) surfaces. *Nature Nanotechnology*, 14, 1, 2019.
15. Adamska, L. & Sharifzadeh, S., Fine-tuning the optoelectronic properties of freestanding borophene by strain. *ACS Omega*, 2, 11, 2017.
16. Ranjan, P. et al., Freestanding borophene and its hybrids. *Advanced Materials*, 31, 27, 2019.
17. Adamska, L., Sadasivam, S., Foley, J. J., Darancet, P. & Sharifzadeh, S., First-principles investigation of borophene as a monolayer transparent conductor. *The Journal of Physical Chemistry C*, 122, 7, 2018.
18. Peng, B. et al., The electronic, optical, and thermodynamic properties of borophene from first-principles calculations. *Journal of Materials Chemistry C*, 4, 16, 2016.
19. Xie, Q.-X. & Zhao, Y., The stability analysis of the monolayer triangular borophene adsorbed on substrates: first-principles simulation. *Computational Materials Science*, 190, 2021.
20. Zhang, Z., Penev, E. S. & Yakobson, B. I., Two-dimensional boron: structures, properties and applications. *Chemical Society Reviews*, 46, 22, 2017.
21. Mortazavi, B., Rahaman, O., Dianat, A. & Rabczuk, T., Mechanical responses of borophene sheets: a first-principles study. *Physical Chemistry Chemical Physics*, 18, 39, 2016.
22. Zhong, H., Huang, K., Yu, G. & Yuan, S., Electronic and mechanical properties of few-layer borophene. *Physical Review B*, 98, 5, 2018.
23. Mortazavi, B. et al., Borophene hydride: a stiff 2D material with high thermal conductivity and attractive optical and electronic properties. *Nanoscale*, 10, 8, 2018.
24. Zhang, L. et al., Recent advances in hybridization, doping, and functionalization of 2D xenes. *Advanced Functional Materials*, 31, 1, 2021.
25. Liang, T., Zhang, P., Yuan, P., Zhai, S. & Yang, D., A molecular dynamics study on the thermal conductivities of single- and multi-layer two-dimensional borophene. *Nano Futures*, 3, 1, 2019.
26. Kou, L. et al., High-mobility anisotropic transport in few-layer γ-B28 films. *Nanoscale*, 8, 48, 2016.

27. Lherbier, A., Botello-Méndez, A. R. & Charlier, J.-C., Electronic and optical properties of pristine and oxidized borophene. *2D Materials*, 3, 4, 2016.
28. Kumar, P. et al., Alpha Lead Oxide (α-PbO): a new 2D material with visible light sensitivity. *Small*, 14, 12, 2018.
29. Feng, B. et al., Discovery of 2D anisotropic dirac cones. *Advanced Materials*, 30, 2, 2018.
30. Xiao, R. C. et al., Enhanced superconductivity by strain and carrier-doping in borophene: a first principles prediction. *Applied Physics Letters*, 109, 12, 2016.
31. Zhao, Y. S., Wu, J. & Huang, J., Vertical organic nanowire arrays: controlled synthesis and chemical sensors. *Journal of the American Chemical Society*, 131, 9, 2009.
32. Zhai, H.-J. et al., Observation of an all-boron fullerene. *Nature Chemistry*, 6, 8, 2014.
33. Tang, H. & Ismail-Beigi, S., Self-doping in boron sheets from first principles: a route to structural design of metal boride nanostructures. *Physical Review B*, 80, 13, 2009.
34. Li, Q. et al., Synthesis of borophane polymorphs through hydrogenation of borophene. *Science*, 371, 6534, 2021.
35. Liu, G., Wang, H., Gao, Y., Zhou, J. & Wang, H., Anisotropic intrinsic lattice thermal conductivity of borophane from first-principles calculations. *Physical Chemistry Chemical Physics*, 19, 4, 2017.
36. Wang, Z., Lü, T.-Y., Wang, H.-Q., Feng, Y. P. & Zheng, J.-C., High anisotropy of fully hydrogenated borophene. *Physical Chemistry Chemical Physics*, 18, 46, 2016.
37. Zhao, L. et al., Mechanistic origin of the high performance of Yolk@Shell Bi2S3@N-Doped carbon nanowire electrodes. *ACS Nano*, 12, 12, 2018.
38. Tsafack, T. & Yakobson, B. I., Thermomechanical analysis of two-dimensional boron monolayers. *Physical Review B*, 93, 16, 2016.
39. Mortazavi, B., Le, M.-Q., Rabczuk, T. & Pereira, L. F. C., Anomalous strain effect on the thermal conductivity of borophene: a reactive molecular dynamics study. *Physica E: Low-dimensional Systems and Nanostructures*, 93, 2017.
40. Sun, H., Li, Q. & Wan, X. G., First-principles study of thermal properties of borophene. *Physical Chemistry Chemical Physics*, 18, 22, 2016.
41. Boustani, I., Systematic LSD investigation on cationic boron clusters: Bn (n = 2–14). *International Journal of Quantum Chemistry*, 52, 4, 1994.
42. Wu, X. et al., Two-dimensional boron monolayer sheets. *ACS Nano*, 6, 8, 2012.
43. Xu, S.-G. et al., Two-dimensional semiconducting boron monolayers. *Journal of the American Chemical Society*, 139, 48, 2017.
44. Liu, X., Zhang, Z., Wang, L., Yakobson, B. I. & Hersam, M. C., Intermixing and periodic self-assembly of borophene line defects. *Nature Materials*, 17, 9, 2018.
45. Kiraly, B. et al., Borophene synthesis on Au(111). *ACS Nano*, 13, 4, 2019.
46. Ma, C. et al., Broadband nonlinear photonics in few-layer borophene. *Small*, 17, 7, 2021.
47. Hou, C. et al., Ultrastable crystalline semiconducting hydrogenated borophene. *Angewandte Chemie - International Edition*, 59, 27, 2020.
48. Boroun, M., Abdolhosseini, S. & Pourfath, M., Separated and intermixed phases of borophene as anode material for lithium-Ion batteries. *Journal of Physics D: Applied Physics*, 52, 24, 2019.
49. Liu, J., Zhang, C., Xu, L. & Ju, S., Borophene as a promising anode material for sodium-ion batteries with high capacity and high rate capability using DFT. *RSC Advances*, 8, 32, 2018.
50. Wang, Z.-Q., Lü, T.-Y., Wang, H.-Q., Feng, Y. P. & Zheng, J.-C., Review of borophene and its potential applications. *Frontiers of Physics*, 14, 3, 2019.

51. Ji, X. et al., A novel top-down synthesis of ultrathin 2D boron nanosheets for multimodal imaging-guided cancer therapy. *Advanced Materials*, 30, 36, 2018.

52. Bhise, K., Kashaw, S. K., Sau, S. & Iyer, A. K., Nanostructured lipid carriers employing polyphenols as promising anticancer agents: quality by design (QbD) approach. *International Journal of Pharmaceutics*, 526, 1, 2017.

53. Fang, J., Nakamura, H. & Maeda, H., The EPR effect: unique features of tumor blood vessels for drug delivery, factors involved, and limitations and augmentation of the effect. *Advanced Drug Delivery Reviews*, 63, 3, 2011.

54. Shi, L., Zhao, T., Xu, A. & Xu, J., Ab initio prediction of borophene as an extraordinary anode material exhibiting ultrafast directional sodium diffusion for sodium-based batteries. *Science Bulletin*, 61, 14, 2016.

55. Ariga, K. et al., Nanoarchitectonics beyond self-assembly: challenges to create biolike hierarchic organization. *Angewandte Chemie - International Edition*, 59, 36, 2020.

56. Li, B. L. et al., Engineered functionalized 2D nanoarchitectures for stimuli-responsive drug delivery. *Materials Horizons*, 7, 2, 2020.

57. Li, L., Zhang, H. & Cheng, X., The high hydrogen storage capacities of Li-decorated borophene. *Computational Materials Science*, 137, 2017.

58. Xie, Z. et al., Two-dimensional borophene: properties, fabrication, and promising applications. *Research*, 2624617,2020.

59. Shukla, V., Wärnå, J., Jena, N. K., Grigoriev, A. & Ahuja, R., Toward the realization of 2D borophene based gas sensor. *The Journal of Physical Chemistry C*, 121, 48, 2017.

60. Cockcroft, D. W., Hoeppner, V. H. & Dolovtch, J., Occupational asthma caused by cedar urea formaldehyde particle board. *Chest*, 82, 1, 1982.

61. Tatullo, M., Gentile, S., Paduano, F., Santacroce, L. & Marrelli, M., Crosstalk between oral and general health status in e-smokers. *Medicine*, 95, 49, 2016.

62. Shang, J., Ma, Y., Gu, Y. & Kou, L., Two dimensional boron nanosheets: synthesis, properties and applications. *Physical Chemistry Chemical Physics*, 20, 46, 2018.

63. Tang, X., Gu, Y. & Kou, L., Theoretical investigation of calcium-decorated β12 boron sheet for hydrogen storage. *Chemical Physics Letters*, 695, 2018.

64. Haldar, S., Mukherjee, S. & Singh, C. V., Hydrogen storage in Li, Na and Ca decorated and defective borophene: a first principles study. *RSC Advances*, 8, 37, 2018.

65. Li, J., Zhang, H. & Yang, G., Ultrahigh-capacity molecular hydrogen storage of a lithium-decorated boron monolayer. *The Journal of Physical Chemistry C*, 119, 34, 2015.

66. Liu, T. et al., Li-Decorated β12-borophene as potential candidates for hydrogen storage: a first-principle study. *Materials*, 10, 12, 2017.

67. Er, S., de Wijs, G. A. & Brocks, G., DFT study of planar boron sheets: a new template for hydrogen storage. *The Journal of Physical Chemistry C*, 113, 43, 2009.

68. Gao, X., Cui, Y., Levenson, R. M., Chung, L. W. K. & Nie, S., In vivo cancer targeting and imaging with semiconductor quantum dots. *Nature Biotechnology*, 22, 8, 2004.

69. Medintz, I. L., Uyeda, H. T., Goldman, E. R. & Mattoussi, H., Quantum dot bioconjugates for imaging, labelling and sensing. *Nature Materials*, 4, 6, 2005.

70. Huang, X., El-Sayed, I. H., Qian, W. & El-Sayed, M. A., Cancer cell imaging and photothermal therapy in the near-infrared region by using gold nanorods. *Journal of the American Chemical Society*, 128, 6, 2006.

71. Beard, P., Biomedical photoacoustic imaging. *Interface Focus*, 1, 4, 2011.

72. Kamal, C., Chakrabarti, A. & Ezawa, M., Aluminene as highly hole-doped graphene. *New Journal of Physics*, 17, 8, 2015.

73. Nagarajan, V. & Chandiramouli, R., Investigation on adsorption properties of CO and NO gas molecules on aluminene nanosheet: a density functional application. *Materials Science and Engineering: B*, 229, 2018.
74. Yuan, J., Yu, N., Xue, K. & Miao, X., Stability, electronic and thermodynamic properties of aluminene from first-principles calculations. *Applied Surface Science*, 409, 2017.
75. Nagarajan, V. & Chandiramouli, R., Interaction of alcohols on monolayer stanane nanosheet: a first-principles investigation. *Applied Surface Science*, 419, 2017.
76. Lin, C., Qin, W. & Dong, C., H2S adsorption and decomposition on the gradually reduced α-Fe2O3(001) surface: a DFT study. *Applied Surface Science*, 387, 2016.
77. Guo, D., Shao, B., Li, C. & Ma, Y., Theoretical insight into structure stability, elastic property and carrier mobility of monolayer arsenene under biaxial strains. *Superlattices and Microstructures*, 100, 2016.
78. Jain, A. & McGaughey, A. J. H., Strongly anisotropic in-plane thermal transport in single-layer black phosphorene. *Scientific Reports*, 5, 1, 2015.
79. Efetov, D. K. & Kim, P., Controlling electron-phonon interactions in graphene at ultrahigh carrier densities. *Physical Review Letters*, 105, 25, 2010.
80. Yeoh, K. H., Yoon, T. L., Rusi, Ong, D. S. & Lim, T. L., First-principles studies on the superconductivity of aluminene. *Applied Surface Science*, 445, 2018.
81. Kansara, S., Gupta, S. K., Sonvane, Y., Hussain, T. & Ahuja, R., Theoretical investigation of metallic nanolayers for charge-storage applications. *ACS Applied Energy Materials*, 1, 7, 2018.
82. Villagracia, A. R. et al., Hydrogen adsorption on calcium, potassium, and magnesium-decorations aluminene using density functional theory. *International Journal of Hydrogen Energy*, 46, 31, 2021.
83. Wundrack, S. et al., Liquid metal intercalation of epitaxial graphene: large-area gallenene layer fabrication through gallium self-propagation at ambient conditions. *Physical Review Materials*, 5, 2, 2021.
84. Metin, D. Z., Hammerschmidt, L. & Gaston, N., How robust is the metallicity of two dimensional gallium? *Physical Chemistry Chemical Physics*, 20, 43, 2018.
85. Singh, D., Gupta, S. K., Lukačević, I. & Sonvane, Y., Indiene 2D monolayer: a new nanoelectronic material. *RSC Advances*, 6, 10, 2016.
86. Kochat, V. et al., Atomically thin gallium layers from solid-melt exfoliation. *Science Advances*, 4, 3, 2018.
87. Singh, D. et al., Effect of electric field on optoelectronic properties of indiene monolayer for photoelectric nanodevices. *Scientific Reports*, 9, 1, 2019.
88. Balandin, A. A., Thermal properties of graphene and nanostructured carbon materials. *Nature Materials*, 10, 8, 2011.
89. Hu, M., Zhang, X. & Poulikakos, D., Anomalous thermal response of silicene to uniaxial stretching. *Physical Review B*, 87, 19, 2013.
90. Gu, X. & Yang, R., First-principles prediction of phononic thermal conductivity of silicene: a comparison with graphene. *Journal of Applied Physics*, 117, 2, 2015.
91. Nissimagoudar, A. S., Manjanath, A. & Singh, A. K., Diffusive nature of thermal transport in stanene. *Physical Chemistry Chemical Physics*, 18, 21, 2016.
92. Losurdo, M., Suvorova, A., Rubanov, S., Hingerl, K. & Brown, A. S., Thermally stable coexistence of liquid and solid phases in gallium nanoparticles. *Nature Materials*, 15, 9, 2016.
93. Bouhafs, D., Moussi, A., Chikouche, A. & Ruiz, J. M., Design and simulation of antireflection coating systems for optoelectronic devices: application to silicon solar cells. *Solar Energy Materials and Solar Cells*, 52, 1, 1998.

94. Nakamura, T., Fujii, H., Juni, N. & Tsutsumi, N., Enhanced coupling of light from organic electroluminescent device using diffusive particle dispersed high refractive index resin substrate. *Optical Review*, 13, 2, 2006.

95. Ou, M. et al., The emergence and evolution of borophene. *Advanced Science*, 8, 12, 2021.

96. Tai, G. et al., Synthesis of atomically thin boron films on copper foils. *Angewandte Chemie - International Edition*, 54, 51, 2015.

97. Mannix, A. J. et al., Synthesis of borophenes: anisotropic, two-dimensional boron polymorphs. *Science*, 350, 6267, 2015.

98. Feng, B. et al., Experimental realization of two-dimensional boron sheets. *Nature Chemistry*, 8, 6, 2016.

99. Li, H. et al., Scalable production of few-layer boron sheets by liquid-phase exfoliation and their superior supercapacitive performance. *ACS Nano*, 12, 2, 2018.

100. Ji, X. et al., A novel top-down synthesis of ultrathin 2d boron nanosheets for multimodal imaging-guided cancer therapy. *Advanced Materials*, 30, 36, 2018.

101. Khalil, B. A. & Gaston, N., Two-dimensional aluminium, gallium, and indium metallic crystals by first-principles design. *Journal of Physics: Condensed Matter*, 33, 12, 2021.

102. Gutiérrez, Y. et al., Polymorphic gallium for active resonance tuning in photonic nanostructures: from bulk gallium to two-dimensional (2D) gallenene %J Nanophotonics. *Nanophotonics*, 9, 14, 2020.

103. Kochat, V. et al., Atomically thin gallium layers from solid-melt exfoliation. *Science Advances*, 4, 3, 2018.

3 Group IVA of 2D Xenes materials (Silicene, Germanene, Stanene, Plumbene)

Yundan Liu, Dan Mu, and Jincheng Zhuang

CONTENTS

DOI: 10.1201/9781003207122-3

3.1 INTRODUCTION

Graphene was firstly obtained in experiment by mechanical exfoliation in 2004, which verified a typical stable freestanding 2D material.[1] Graphene possesses unique properties in optical, electrical, and mechanical applications, which have attracted widespread attention in the study of materials, physics, chemistry, and semiconductors.[2–4] At the same time, the discovery of graphene has also opened the new research era on exploration and application of other new two-dimensional materials. In the periodic table, silicon, germanium, stannum, and lead are in the same main group of carbon (group IV). They have a similar valence electron configuration to carbon, and their 2D allotropes are called silicene, germanene, stanene, and plumene, respectively.[5] Graphene is composed of a monoatomic layer of covalently bonded sp^2 hybridized carbon atoms arranged in a hexagonal honeycomb lattice structure.[1] However, silicon, germanium, stannum, and lead atoms tend to adopt a mixed degree of sp^2–sp^3 hybridization-formed silicene, germanene, stanene, and plumene with buckled honeycomb structure. Motivated by recent advances in the exciting class of group IV Xenes, herein we review the literature to date and connect the structure, properties, synthesis, and application of these elemental 2D materials.

3.2 SILICENE

3.2.1 ATOMIC STRUCTURES OF SILICENE

As early as 1994, Takeda and Shiraishi tried to theoretically search for equivalents similar to graphite silicon and germanium by first-principles total-energy calculations. The results indicated 2D silicon and germanium layers, the equivalent structure of silicon and germanium, prefer to form the D_{3d} corrugated structure, rather than forming the D_{6h} flat structure as demonstrated in the structure of the 2D graphite carbon layer.[6] After the birth of graphene, Guzman-Verri and Lew Yan Voon named graphene-like silicon 2D nanostructure as "silicene" and revealed its electronic properties through tight-binding Hamiltonian theory. The results showed that silicene is a metal or zero-gap semiconductor. The existence of Dirac cones in band structure may lead to the electrons near the K point behaving as Dirac massless fermions with the Fermi velocity estimated at 10^5 m/s.[7] In 2009, through systematic calculations, Cahangirov et al. theoretically verified that low-buckled silicene can exist stably. Figure 3.1a–b illustrates their calculated results of the phonon modes and the function relationship between the binding energy and lattice constant of silicene in forms of planar, low-buckling, and high-buckling. Among these three silicene structures, the planar structure of silicene was an unstable configuration because the minimum binding energy is larger than the other two structures and there are imaginary frequencies in the phonon spectrum. In contrast, the low-buckled silicene structure was dynamically stable due to there being no imaginary frequencies in the phonon spectrum. The dynamic stability of the low-buckled silicene was further confirmed by ab initio finite temperature

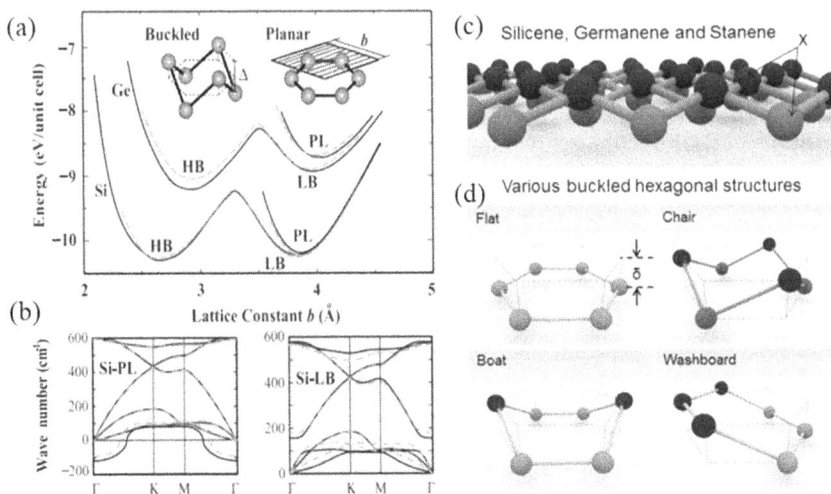

FIGURE 3.1 (a) Energy versus hexagonal lattice constant of 2D Si and Ge are calculated for various honeycomb structures. Planar and buckled geometries together with buckling distance and lattice constant of the hexagonal primitive unit cell. (b) Phonon dispersion curves obtained by force-constant and linear response theory, respectively. Reproduced with permission.[8] Copyright 2009, APS. (c) Buckled hexagonal crystal structure of 2D elemental sheets (X = Si, Ge, and Sn). (d) Various hexagonal buckled structures of X. The buckling parameter δ is the vertical distance between the two planes of X atoms. Reproduced with permission.[12] Copyright 2015, Wiley.

molecular dynamics calculations, and the structure is still undamaged even if the temperature raised to 1000 K.[8] Actually, the obtained melting point of silicene is 1750 K by using Monte Carlo simulations, which is much lower than the melting point of bulk silicon calculated under the same method.[9]

The s orbit in carbon atoms hybridizes one p orbital at a time due to the relatively large energy cost, so that they prefer to form linear sp atomic chains or 2D planar sp^2 configurations with honeycomb structures. However, in silicon, the energy of hybridizing s and p orbitals is much smaller than (approximately half of) that in carbon. Thus, it is feasible to hybridize all the three p orbitals with s orbitals in silicon, making a sp^3 three-dimensional(3D) tetrahedral configuration is a more preferred structure. In addition, Si has a larger ionic radius than C and the lengths of the Si–Si bond are much larger than that of the C–C bond, which results in the overlapping of the perpendicular p_z orbital and hence making the p bond formation more difficult in silicon. As indicated in the report, the lattice constant of the low-buckled silicene is 3.83 Å, where the nearest neighboring distance of Si is 2.25 Å and the height of buckling is 0.44 Å.[3,10,11]

The hexagonal structure with a buckled nature of silicene is illustrated in details in Figure 3.1c, where some atoms of the unit cell are in a lower atomic plane and some are at upper position. Each Si atom is covalently bonded with three others, constructing a simple hexagonal unit cell. As shown in Figure 3.1d, different types of buckling lattice arrangements such as chair-, boat-, and washboard-like structures are possible in principle, which may occur at varying points in the unit cell. Nevertheless, DFT analysis revealed that the boat- and washboard-like arrangements are not stable enough, which may evolve into a flat structure. The chair-like structure is more stable, even in comparison to the flat structure. The specific buckling parameter (δ) is the vertical distance separating the two atomic planes in these structures, which is used to refer to the variation of buckling structures. The regulation on the buckling parameter could directly regulate the properties of silicene, such as the density of electronic states.[12,13]

3.2.2 Basic Properties of Silicene

In silicene, the buckled Si atoms make themselves closer for a stronger overlapping, resulting in a hybridized sp^2–sp^3 configuration with a lower symmetry. The spin–orbit coupling (SOC) effects would be more intensive due to the heavier mass of silicon, which may induce the topologically nontrivial electronic states. The buckled honeycomb structure and enhanced SOC effect offer silicene unique electronic properties which are different from graphene. These properties, including tunable bandgap structure, spin-polarized edge states, and topologically nontrivial electronic states, indicate a great potential for application in nano-electronics and spintronics.[14]

Without considering the SOC, the band structure of the freestanding silicene based on density functional theory (DFT) calculation is shown in Figure 3.2a.[15] The results displays a linear Dirac-type dispersion of electron state with a zero-gap near the K and K' points, which is similar with that of graphene. Considering SOC effect, a spin-orbit gap of about 1.4–2.0 meV could be observed in the calculated band structure of silicene.[9] Another report, derived from quasi particle interference patterns, indicated the energy dispersion in silicene is linear in momentum which is coupled with a large Fermi velocity (1.2×10^6 ms^{-1}), as shown in Figure 3.2b. It revealed the existence of the Dirac cone in the electronic band structure of low-buckled silicene.[8,16,17] Experimental evidence also supports that Fermi velocities of silicene overlaid on Ag substrate are comparable to that of graphene, and the nature of the quasi particles in silicene behave similarly to the massless Dirac fermions.[11,18] It should be noted that in such cases the effect of underlying substrate could be excluded in order to determine the intrinsic properties of silicene. A recent work realized the oxidization of bilayer silicene on Ag (111), and the top layer of silicene exhibits the signature of the 1 × 1 lattice structure of "freestanding" silicene which displays a robust Dirac fermion characteristic with less electron doping.[19]

In addition to the influence of substrates, the bandgap structure of silicene can also be regulated via applying strain, an external electric field, and the oxidation effect. As shown in Figure 3.2c, Liu et al. have theoretically shown that

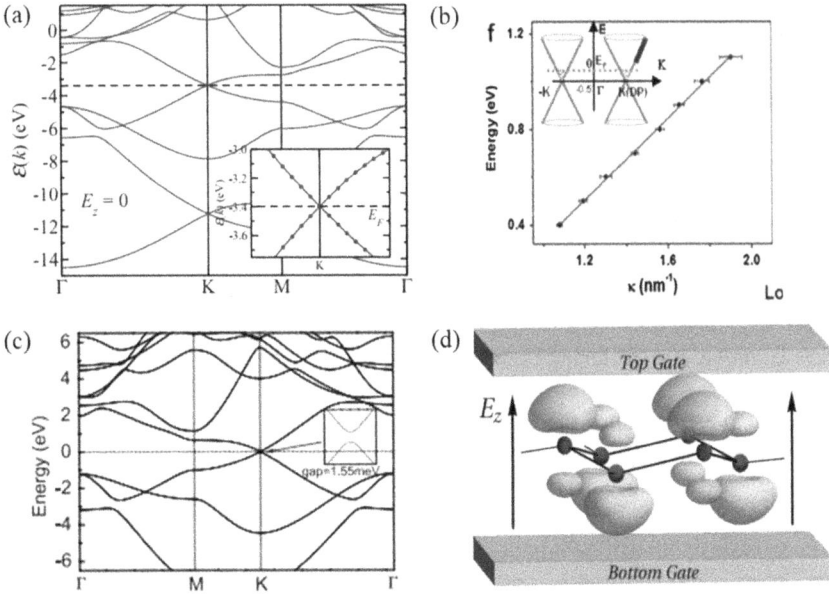

FIGURE 3.2 (a) Band structure of the freestanding silicene without the consideration of SOC and external electric field based on DFT calculations. Reproduced with permission.[15] Copyright 2012, APS. (b) Energy dispersion as a function of κ for silicene determined from the wavelength of QPI patterns. Reproduced with permission.[16] Copyright 2012, APS. (c) Band structure of low-buckled silicene. Reproduced with permission.[20] Copyright 2011, APS. (d) Atomic structure of silicene, together with a sketch of the charge density for the highest occupied valence band in the vicinity of the K point. Reproduced with permission.[15] Copyright 2012, APS.

low-buckled honeycomb geometry silicene can open a band gap of 1.55 meV with SOC at K point. The direct calculation of the Z_2 topological invariant shows that low-buckled silicene featured with topologically nontrivial electronic structures can induce quantum spin Hall effect (QSHE). Further, they found that the band gap can be adjusted to 2.9 meV by applying −6% strain.[20] As demonstrated in Figure 3.2d, the sublattice symmetry and the band structure of silicene were modulated by applying an external electric field along the z-direction. The calculated results indicate that the symmetry structure of the buckled honeycomb silicene could be broken. Then a gap could be opened in the band structure, and the gap size can be established as a function of the electric field strength. Ni et al. further regulated the electronic properties of free-standing and h-BN-sandwiched silicene by applying a vertical electric field, and then revealed it has a linear influence on the bandgap.[14, 21] Due to the high reactivity of silicon, silicene is unstable under ambient conditions. Chemical functionalization is an attractive strategy to stabilize silicene in air, such as hydrogenation, oxidization, or chlorination of monolayer silicene.[14]

3.2.3 Preparation of Silicene

3.2.3.1 Epitaxial Growth of Silicene

After the discovery of graphene, the structure of silicon has received widespread attention. However, silicon is different from carbon, which has a natural element of only sp^3 hybridized diamond structure. Therefore, the exfoliation method similar to graphene cannot directly be used with silicene. Learning from the synthetic growth of graphene, molecular beam epitaxy has become a common method. In 2012, several research groups reported the results of the successful preparation of monolayer silicene at the same time. Among them, the silicene grown on the Ag (111) surface attracts the most attention.[22,23] When the Ag (111) substrate temperature is above 400 K during deposition, the silicene sheet exhibits typically four ordered phases with increasing substrate temperature and coverages as shown in Figure 3.3a–d. Because the lattice constant of Ag (111) (0.288 nm) is quite different from that of free suspended silicene (0.383 nm), the monolayer honeycomb silicene grown on the surface of Ag (111) can be rotated at an angle relative to the silver to achieve reconstruction. However, different stacking methods can make silicon atoms with different phases directly above silver atoms, resulting in different out-of-plane warping reconstruction patterns. Therefore, the silicene growth based on silver substrate shows a variety of different phase reconstruction.

Besides the most frequently used Ag (111), other metal substrates with moderate interfacial interactions and matching lattice constants are also used to grow silicene Considering that silicene was originally epitaxy on the surface of silver, it is logical that silicene can also grow on the surface of gold. Physically speaking, gold behaves the same as silicon. Silicon nanoribbons on Au (110) have been successful reported as shown in Figure 3.3e–f. The results show that the contact between the ordered Au-Si alloy surface layer and silicon will lead to the growth of parallel assembled straight silicon nanoribbons. Limited by the width of the Au (110) terrace, the nanoribbons have the same orientation, the main characteristic narrow width is 1.6 nm, and the length can reach several hundred nanometers.[24] Another metal substrate used for the growth of silicene is iridium, which has received extensive attention recently. Silicene sheets can be epitaxially grown on Ir (111), which matches the $(\sqrt{7} \times \sqrt{7})$ superstructure of Ir (111) lattice and is consistent with the $(\sqrt{3} \times \sqrt{3})$ superlattice of silicene. The difference in the position of different silicon atoms relative to the crystal lattice of the substrate is the main reason for the buckling conformation of the silicene sheet in Figure 3.3g.[25] As shown in Figure 3.3h, the monolayer silicene grown on Ru (0001) has a similar structure to that grown on Ir (111), such as the $(\sqrt{7} \times \sqrt{7})$ superstructure. This feature may be related to the similar lattice constant (0.27 nm) of Ru and Ir substrates. This work identified the unique monolayer structure of silicon grown on Ru (0001) substrates and clarified the step-by-step growth of silicene.[26]

So far, only the metal surface has been studied as a suitable substrate for the growth of silicene or nanoribbons. Now we discuss the growth of silicene or silicon layers in non-metal crystalline materials. Zirconium diboride can be epitaxially grown into crystals. Fleurence et al. demonstrated that epitaxial 2D

FIGURE 3.3 Silicene on Ag(111): phase transition and other substrates. (a) $\sqrt{13}\times\sqrt{13}$; (b) 4×4; (c) $2\sqrt{3}\times2\sqrt{3}$; (d) $\sqrt{3}\times\sqrt{3}$. Reproduced with permission.[22] Copyright 2012, ACS. (e) STM image of the surface at 0.3 Si ML coverage showing nanoribbons oriented along the $(11\bar{0})$ direction. (f) High resolution STM image showing the structure of the 1.6 nm wide Si nanoribbons. Reproduced with permission.[24] Copyright 2013, AIP. (g) STM image of the (7×7) superstructure on Ir (111) surface. Reproduced with permission.[25] Copyright 2013, ACS. (h) STM image of ($\sqrt{7}\times\sqrt{7}$) superstructure on Ru (0001). Reproduced with permission.[26] Copyright 2017, ACS. (i) STM image of the (2×2)-reconstructed ZrB$_2$ (0001) surface. Reproduced with permission.[27] Copyright 2012, APS. (j) STM image of the epitaxial Si nanosheet on MoS$_2$. Reproduced with permission.[29] Copyright 2012, Wiley.

silicene is spontaneously formed by surface segregation on a zirconium diboride film grown on Si (111) wafers. The large-scale STM image of the (2×2) reconstructed ZrB$_2$ (0001) surface is striped as shown in Figure 3.3i. The driving force behind the mechanism that triggers the appearance of fringe patterns was found to be related to phonon instability, which was characterized by the zero-frequency mode reaching the critical point of instability. Through low-temperature STM and STS, Florence et al. also proved the n-type semiconductor properties of epitaxial silicene on ZrB$_2$ (0001).[27,28] Using MBE technology, 2D Si nanosheets were successfully fabricated on MoS$_2$, which is a semiconducting substrate. The STM image in Figure 3.3j demonstrates the presence of Si nanosheets partially covering the surface of MoS$_2$.[29] Both STS and DFT calculations showed that the obtained high buckling silicene is metallic. Although the silicene sheet on the surface of

MoS_2 was severely distorted, this is the first step in the synthesis of silicene on non-metallic substrates and is the key to future silicene nanoelectronic devices.[29]

However, no matter what kind of silicene phase, theoretical predictions of the Spin Hall effect have not been observed experimentally and more methods need to be tried. Researchers reported an experimental study of the chlorination reaction of a monoatomic silicene layer on Ag (111). The study confirmed that silicene can form an ordered 1×1 structure through chlorination, which provides a platform for novel quantum phenomena such as the QSHE predicted by subsequent research theories. The adsorption of a layer of chlorine on the surface may be able to better protect the silicene from oxidation.[30] At the same time, other methods of separating silicene from the substrate have also been reported. Du et al. developed a quasi-freestanding silicene method based on intercalation. The oxygen atoms are inserted into the lower layer of silicene, causing the upper layer of silicene to be isolated from the substrate. Therefore, the top layer of silicene exhibits the characteristics of a 1×1 honeycomb lattice and the oxygen insertion preserves the properties of massless Dirac fermions.[19]

Silicene has a more flexible crystal structure than graphene, allowing for band structure engineering by changing epitaxial conditions. These experiments provide a platform for novel quantum phenomena such as the QSHE predicted by subsequent research theories.

3.2.3.2 Chemical Exfoliation of Silicene

Although there is no natural layered Si bulk for directly exfoliation, the topological chemical deintercalation process can be applied to possible synthesis of 2D silicene, for example, by selectively removing calcium from the precursor Zintl phase silicide (such as $CaSi_2$). Researchers reported the synthesis of scalable, independent, few-layer silicene nanosheets through liquid oxidation and $CaSi_2$ exfoliation. The prepared silicene nanosheets are single-layer or thin dispersible ultra-thin super-flake with good crystallinity, and they have a hexagonal pattern structure similar to silicene as shown in Figure 3.4a.[31] A similar top-down strategy, electrochemical exfoliation, may be available for synthesis of silicene. Zhang

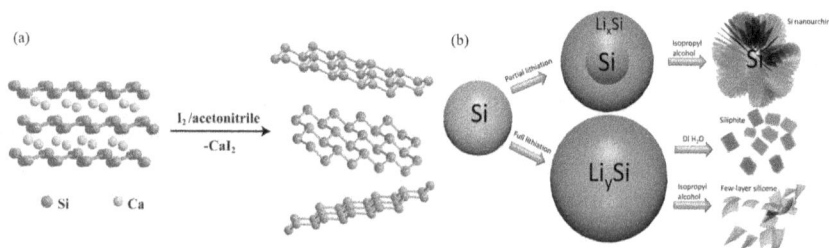

FIGURE 3.4 (a) Schematic illustration for the synthesis of silicene from $CaSi_2$ via liquid oxidation and exfoliation. Reproduced with permission.[31] Copyright 2018, Wiley. (b) Scheme of lithiation and delithiation process of silicon in different scenarios. Reproduced with permission.[32] Copyright 2018, Wiley.

et al. reported several layers of silicene nanosheets produced by continuous lithiation and decolorization using selected protic solvents as shown in Figure 3.4b. Studies have found that denitrification solvents play a vital role in the evolution of the final product structure. The top-down synthesis strategy proposed in this study not only provides a new solution to the problem of low-layer silicene preparation, but also proves the feasibility of preparing 2D materials from non-layered starting structures.[32] These different preparation methods have paved the way for the development of silicene in the future and broadened the scope of application of silicene. Combining theoretical modeling and calculations to gain an in-depth understanding of the formation mechanism, the research on large-scale synthesis of silicene would arouse more interest of researchers.

3.2.4 APPLICATION OF SILICENE

Until now, theoretical and experimental efforts have been made to explore the applications of silicene, including batteries, field effect transistors (FET), gas sensors, etc.

3.2.4.1 Batteries

For battery application, Si-based anodes have already been intensively explored due to their high capacity, low operation potential, and that they are economic and pollution-free. One biggest challenge for their practical implementation is the large volume variation during lithiation and de-lithiation processes which results in unstable performance and safety issues.[33] Using a unique 2D layered silicon allotrope, i.e., silicene, may effectively solve this problem. Tritsaris et al. applied the DFT computational method to evaluate the lithium-ion storage capability of single-layer and few-layer (double-layer) silicene. The theoretical results indicate the specific charge capacity of silicene to be 954 and 715 mAh/g for the single-layer and double-layer, respectively.[34] The experimental results also demonstrate that the anode made of silicene nanosheets exhibited a nearly theoretical capacity of 721 mAh•g^{-1} at 0.1 A•g^{-1} (Figure 3.5a) and an extraordinary cycling stability with no capacity decay after 1800 cycles.[35] In addition, through modifying the surface of silicene with saturated -H and -OH, the as-prepared siloxane can deliver a reversible capacity of 2300, 311, and 203 mAh/g for Li, Na, and K, respectively. In contrast to the rapid capacity decay of previous reported silicon anodes, the durable performance of silicene as an anode material offers silicene a great potential for energy storage and beyond.[36,37]

3.2.4.2 Field Effect Transistor

The theoretical high carrier mobility, the possible QSH effect, and the natural compatibility with mature silicon technology makes silicene quite prominent as the channel material of FET devices with high-speed and good gate control. Many works have simulated and predicted the excellent and promising performances of silicene towards FETs. Ni et al. theoretically showed that it is possible to open a bandgap in a semi-metallic buckled silicene monolayer by applying a

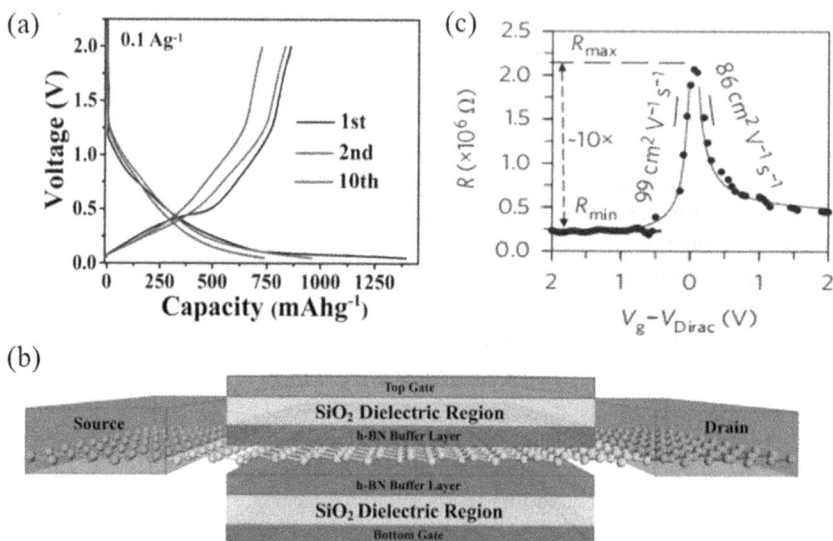

FIGURE 3.5 (a) The 1st, 2nd, and 10th charge–discharge curves of silicene nanosheets anode at a current density of 0.1 A g^{-1} within a voltage range from 0.05 to 2.0 V versus Li+/Li. Reproduced with permission.[35] Copyright 2018, Wiley. (b) The schematic model of Dual-gated silicene FET. Reproduced with permission.[21] Copyright 2012, ACS. (c) The carrier mobilities extracted from the R versus (V_g-V_{dirac}) plot of silicene FET. Reproduced with permission.[39] Copyright 2015, *Nature*.

perpendicular applied electric field (Figure 3.5b).[21] The bandgap size can be regulated linearly by means of electric field strength. However, due to the stability and synthesis portability of silicene, the experimental progress on realizing silicene FET fell behind the theoretical investigations.

Tao et al. successfully fabricated the silicene FET by transfer and integration for the first time. As demonstrated, the silicene-based FETs have been successfully fabricated by transferring the grown silicene from Ag (111)/mica substrate to the SiO$_2$/Si. The silicene device demonstrated in this work agrees well with the predictions of Dirac-like ambipolar charge transport. However, the measured carrier mobility of the as-prepared silicene is ~100 cm$^2 \cdot$V$^{-1} \cdot$s^{-1}, which is only 0.1% of the theoretical value (Figure 3.5c). Improving the performance of silicene-based electronic devices, especially on the band structure optimization and carriers' transport enhancement, is still the key challenge.[38–40]

3.2.4.3 Gas Sensors

The peculiar electronic properties of silicene enable the capability to sense certain gaseous molecule and chemical species, which offers the potential of silicene for chemical sensor application. Applying DFT calculations, Feng et al. systematically investigated the adsorption of gas molecules including CO, NO, NO$_2$, O$_2$, CO$_2$, NH$_3$, and SO$_2$ on silicene sheet. Silicene is regarded as a promising sensor for NO and NH_3 since these two molecules can be chemically adsorbed on silicene with moderate adsorption energies (0.35 eV for NO and 0.6 eV for NH_3).

In addition to the sensing capability, silicene has been also considered as a promising hydrogen storage material. According to the reference, the weight percent of hydrogen in silicene could range from 6.6% (Si_6H_{12}) to 4.5% ($Si_{70}H_{92}$) after full hydrogenation on both upper/lower sides of silicene as well as modifying the edge-dangling Si atoms.[41] Experiments done by Wu's group demonstrated that, by annealing to a moderate temperature of about 450 K, the fully hydrogenated silicene sheet on Ag (111) substrate is able to transfer to the pristine silicene without hydrogen absorbance.[42] The feasibility of reversible hydrogenation offers silicene a great potential for controllable chemical storage of hydrogen. More and more research is devoted to exploring the physical properties of silicene in order to find more useful materials.

3.3 GERMANENE

3.3.1 Atomic Structures of Germanene

Germanene is also analogous of graphene. Germanene and silicene were theoretically studied almost at the same time, and they have many similarities in structure and properties. The theoretical simulation on the 2D structure of germanene indicated that the 2D honeycomb lattice with buckled configuration is a stable structure with the lowest energy. In the direction perpendicular to the sheet of this buckled configuration, the position of the two honeycomb sub-cells shift slightly relative to each other.[6] Subsequent research calculations show that the phonon spectrum of low-buckled germanene also has no imaginary frequencies in the Brillouin zone, and this structure is still stable when the temperature increases to 800 K. In detail, the relevant lattice parameters, buckling distance, and nearest neighbor distance of low-buckled germanium honeycomb structures are 3.97 Å, 0.64 Å, and 2.38 Å, respectively.[8] Kaloni and Schwingenschlögl reported that, even applied with 16% biaxial tensile strain, the germanene is still stable in their DFT calculation, which reflects the robust of buckled germanene structure.[43]

3.3.2 Basic Properties of Germanene

The electronic structures of germanene are similar with graphene. As a zero-gap semi-metallic, the charge carriers in germanene are massless fermions due to the linear π and π^* bands at the Fermi level.[6] In view of the fact that the intrinsic spin–orbit coupling strength of Ge atoms is greater than C atoms, an intrinsic bandgap could be opened up at the Dirac point. It indicates that the magnitude of the gap induced by effective SOC in a low-buckled structure is 23.9 meV. Accordingly, it is natural to conceive that the QSH effect is more significant in germanene, and the recent reports also demonstrate the realization of QSH effect in germanene. Given the comparative ease on the integration of germanene into the Si-based nanoelectronics, germanene with the QSH effect attracts high levels of interest to construct novel spintronic devices.[20]

Compared with the four-fold coordination in bulk Ge, the surface Ge atoms in germanene are unsaturated and prefer to adsorb foreign atoms or molecules, which indicates the instability of germanene in ambient conditions.[44] Generally, surface modification affects the linear energy spectrum, a band gap may be induced at the Dirac points, and finally the high-speed carriers can be damaged. Therefore, artificially enhancing the germanene's immunity to ambient conditions and maintaining its massless electron character would be highly desirable, which can significantly benefit the application of germanene.

Manipulation on the symmetry between two underlying sublattices of the honeycomb buckled structures enable a way to regulate the properties of germanene. For example, the magnetism of germanene can be induced by breaking the symmetry of sublattice through decorating chemical bonds only on one side of the buckling structure.[45] In the case of saturating chemical bonds, on one side two narrow bands at half filling always exist. In addition, due to the presence of the Hubbard U interaction, a flat band ferromagnetism could be induced and thus magnitude can also be controllable by tuning the saturation fraction.

The structural, electronic, and magnetic properties of germanene can be further regulated by adsorbing TM atoms. It was found that adsorbing Sc, Ti, V, Cr, Mn, Fe, and Co on germanene can induce magnetism, while nonmagnetic semiconducting states could be realized with the absorbance of Ni, Cu, and Zn. The band structure of germanene can be also tuned by adsorbing atoms such as Sc, Ni, Cu, and Zn, where the largest gap can be regulated up to 74 meV. The theoretical results also indicated that it is possible that germanene adsorbed with V atoms can host the quantum anomalous Hall (QAH) effect.[46]

3.3.3 Preparation of Germanene

Germanene with a honeycomb structure was first synthesized on Au (111) surface by MBE in 2014 and the grown $\sqrt{3} \times \sqrt{3}$ (R30°) germanene superstructure matches well with the $\sqrt{7} \times \sqrt{7}$ (R19°) Au (111) supercell.[47] Silver has been also suggested as a suitable substrate for germanene epitaxy.[48,52] Unlike silicene, Ge segregated on Ag (111) would epitaxially grow on the $7\sqrt{7} \times 7\sqrt{7}$ (R19.1°) supercell with respect to Ag (111). As shown in Figure 3.6a, the results show two distinct phases on Ag (111) single crystal by MBE.[48] The formation of germanene has been also successfully grown on other metal surfaces, such as Pt (111),[49] Al (111),[50] Sb (111),[51] Ge (111),[53] Ge$_2$Pt,[54] and Au (111).[55] Some STM images are shown in Figure 3.6. Different degrees of bulking in structure could be induced by the mutual coupling between the germanene and the metal substrate, which then results in different periodic structures. Moreover, hybridization with the electronic states of the metallic substrate usually damage the Dirac nature of 2D germanene. Qin et al. have successfully grown a single layer of germanene on a Cu (111) substrate.[56] As shown in Figure 3.6b, the distance between the nearest

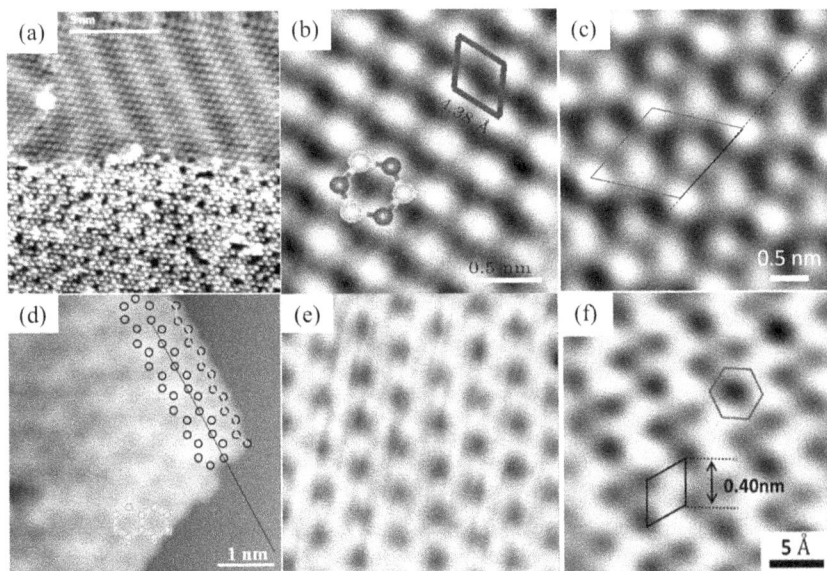

FIGURE 3.6 (a) Coexisting striped phase and quasi-freestanding phase germanene on Ag (111). Reproduced with permission.[48] Copyright 2018, APS. (b) Zoomed-in STM image of a germanium adlayer on Cu (111), revealing the periodicity of the germanene superstructure (4.38 Å). Reproduced with permission.[56] Copyright 2017, Wiley. (c) Zoomed-in STM image of the germanium adlayer on Pt(111). Reproduced with permission.[49] Copyright 2014, Wiley. (d) Topographic STM image of the monolayer germanene on Cu(111). Reproduced with permission.[56] Copyright 2017, Wiley. (e) STM image of the germanene on Al (111).Reproduced with permission.[50] Copyright 2015, ACS. (f) Atomic view of a germanene lattice grown on Ag (111). Reproduced with permission.[52] Copyright 2018, Wiley.

neighbor atoms (about 2.54 Å) of the grown structure is slightly larger than that in the theoretical prediction (about 2.38 Å). However, the Dirac electronic state was not observed, due to the strong interaction between germanene and the Cu (111) substrate.

It should be noted that layered substrates like graphite[57] and MoS_2[58] are also possible for the growth of germanene. Zhang et al. reported the successful synthesis of germanene on molybdenum disulfide (MoS_2). It was found that in the coupling between the germanene layer and MoS_2, the substrate is very weak, so a clear V-shaped density of states can be observed in germanene.[58]

In addition to MBE, other strategies for the fabrication of 2D germanene and hybridized germanene materials have been experimentally demonstrated in recent years. In 2013, Bianco et al. firstly realized the synthesis of a hydrogen-terminated multilayer germanium (GeH). The multilayered GeH bulk was prepared by de-intercalation of $CaGe_2$ in aqueous HCl for more than one week, as demonstrated in Figure 3.7. They have created gram-scale, millimeter-sized crystallites of hydrogen-terminated germanane, which exhibits high thermal

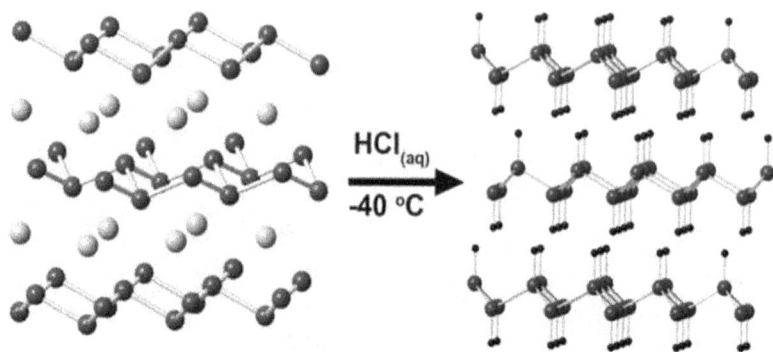

FIGURE 3.7 Schematic illustration of topotactic deintercalation of CaGe₂ to GeH. Reproduced with permission.[59] Copyright 2013, ACS.

stability and long-term resistance to oxidation.[59] Recently, another group has fabricated Ge-based nanostructures evolved from GeH, with controlled morphology and structure as well as chemical composition. This 2D germanene nanosheet architecture with preserved hexagonal structure has been applied in sodium ions batteries, displaying higher reversible and rate capabilities compared with counterparts including GeH nanosheets and Ge particles.[60]

3.3.4 APPLICATON OF GERMANENE

As a Dirac fermion material, germanene is regarded as a promising candidate for electronics such as FET, with high-speed and low-energy consumption. In addition, numerous applications in different fields involving germanene and functionalized germanene have been reported in recent years.[61–64]

3.3.4.1 Batteries

Germanene and functionalized germanene are promising two-dimensional materials. Due to their low cost and high energy density, some researchers have extensively studied them for energy conversion and storage devices in battery applications, such as electric vehicles, portable electronic devices, and large power grids (in lithium-ion batteries).[91] The first-principles DFT calculation of Mortazavi et al. was used to evaluate and analyze the insertion of Li/Na-ions on the 2D germanene film as shown in Figure 3.8a. The most stable binding site and binding energy were found after the insertion of germanene. The results based on Bader charge research show that the energy density of Li/Na ions deposited on germanene nanosheets is about 369 mAh g⁻¹. Theoretical research has opened up a good path for the application of germanene nanosheets in lithium- and sodium-ion batteries.

3.3.4.2 Optoelectronics

Similar with silicene, 2D germanene and hydride germanene with tunable band structure have also obtained great attention for their optoelectronics potential.

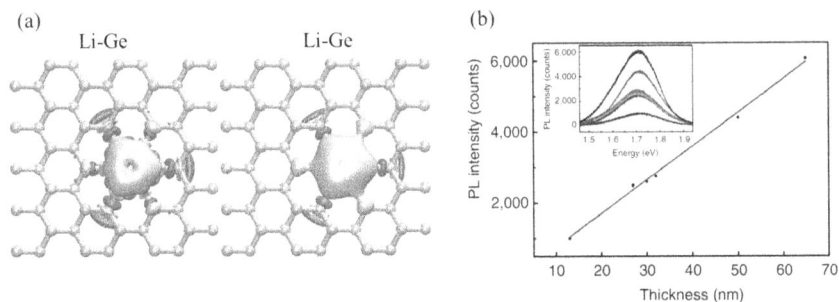

FIGURE 3.8 (a) Calculated charge density difference plots for Li or Na ions on germanene. Reproduced with permission.[91] Copyright 2016, Elsevier. (b) PL intensity of exfoliated GeCH$_3$ thin flakes having average thicknesses ranging from 13–65 nm. Inset is the raw photoluminescence spectra of flakes in the reaction steps. Reproduced with permission.[66] Copyright 2018, *Nature*.

The hydrogenated germanene, i.e., GeH, is not only stable in air, but also possess a direct band gap around 1.6 eV that can respond to the visible light. The electron mobility of GeH can be established at 18000 cm^2 V^{-1} s^{-1}, which is obviously higher than that of crystalline germanium. The good photo-response and high charge transfer ability offer germanene as a promising optoelectronic material.[65] Several prototypes of optoelectronic devices involving GeH as active materials have been successfully fabricated which have exhibited attractive photoelectric conversion efficiency.[66] Replacing the hydrogen in GeH with methyl terminates (–CH$_3$) could further tune the band gap of hydride germanene. A band-edge fluorescence with a quantum yield of ~0.2% , independent of the layer thickness, was found in methyl-terminated germanene. It was found that, in comparison to GeH, the band gap of methyl-terminated germanene is about 1.7 eV, and the thermal stability is also improved (Figure 3.8b). This means that maintaining a high quantum yield of such hydride germanene will not be limited to the monolayer, which can significantly promote the practical application of germanene.

With the rapid development of intensive research on this unique class of 2D nanomaterials, high-purity germanene-based materials are expected to be used in a wide range of electronic, optoelectronic, and thermoelectronic applications.

3.4 STANENE

3.4.1 Atomic Structures of Stanene

Sn has two stable allotropes in the bulk phase: α-Sn (gray tin) which is a less dense diamond cubic structure and β-Sn (white tin) which is a malleable tetragonal structure. As another analogous of graphene, stanene is a monoatomic layer of Sn with a buckled honeycomb-like structure. The primitive cell of stanene contains two atoms with an optimized lattice constant of ≈4.67 Å.[67–70] In the 2D structure of stanine, the large size of Sn atoms induces strong inner-core repulsion forces, which are more intense than that of germanene and silicene.[69] Stanene is also a buckled structure with a more obvious wave-like structure, where the adjacent atoms prefer an out-of-plane orientation and short Sn-Sn bond. Such a corrugated structure is featured with the hybridization of sp^2 and sp^3 orbital states. The theoretical results indicate that, in the quasi-free-standing state, the puckered structure of stanene could be possibly stable in both high-buckled and low-buckled structures. The high-buckled structure has a large difference in the out-of-planar height between neighboring sublattices which exhibit ninefold atomic coordination. This structure is structurally similar to the stable bulk β-Sn. The low-buckled stanene is also a hexagonal close-packed bilayer structure, with an increased lattice constant which is more than 1.2 time than typical parameters (3.42 Å) of high-buckled structure.[71]

The buckling parameter of stanene can be regulated by different external factors, such as the interfacial strain from substrate, applied electric field, and chemical functionalization.[67] Jiang et al. have studied the relationship of bandgap and band energy of stanene monolayer related to the buckling state by a tight binding calculation.[72] Their analysis demonstrated a high cohesive energy of Sn atoms and positive frequencies over all the Brillouin zone, revealing that the hexagonal stanene with a dumbbell geometric configuration is the most stable. The influence of the substrate and external strain on the stable configuration of stanene has been studied through the theoretical route.[67]

3.4.2 Basic Properties of Stanene

Stanene is predicted to have a large spin–orbit band gap of about 70 meV, exhibiting the highest degree of SOC in comparison to silicene and germanene.[73,74] It is also highly expected to realize QSHE in stanene. In addition to the intrinsic SOC effect, the band structure of stanene can be significantly affected by the substrate's interaction and layer stacking arrangements. To avoid the strong interaction of underlying substrate, surface passivation or functionalization of the stanene is utilized to isolate freestanding stanene or preserve the intrinsic properties.[75] For example, Xiong et al. concluded that the electronic properties of both bare and edge-hydrogenated stanene nanoribbons are semiconducting in nature, even if they have different SOC states.[76]

Moreover, applying stain, electric field, and heteroatom modifying are also effective strategies to tune the electronic properties of stanene. Taking stain modulation as the example, Modarresi et al. suggested that the SOC bandgap of stanene can even close by applying a strain.[77] Wang et al. predicted that QSHE could be induced in stanene by applying tensile/compressive strain in the presence of a good lattice match between sample and substrate.[69] Broek et al. suggested that the band gap can be opened up to 0.21 eV by applying lateral strain on stanene and an electric field at the out-of-plane direction.[73,78]

The magnetic characteristic of stanine is another attractive property, which has been theoretically predicted. Xiong et al. predicted that the zigzag stanene NRs are ferromagnetic with opposite spin order between the two edges.[76] Theoretical results also indicated that the adsorption of transition metals on the surface of stanene could regulate the magnetic moment. The magnetic moment of most adsorbed metals' modeling survived; however, stanene adsorbed with Ni, Cu, and Zn is nonmagnetic. Specifically, it was found that the Fe-absorbed stanene is a bipolar magnetic gapless semiconductor containing carriers of up-spin electrons and a down-spin hole. Thus, the charge and pure spin current may coexist, enabling the possibility of electrically controllable spintronic devices.[79] Modeling with stanene NRs, Fu et al. revealed that there are four possible types of edge magnetic configurations, i.e., out-of-plane ferromagnetic, out-of-plane anti-ferromagnetic, in-plane ferromagnetic, and in-plane antiferromagnetic.[80] Spontaneous magnetism was suggested to exist in all these four magnetic states. The recent study also found Rashba-type spin-splitting in stanene gas systems, promising a great potential for spintronics applications.[81]

3.4.3 PREPARATION OF STANENE

In 2013, Xu et al. formally proposed the concept of Tin thin.[82] Later, Barfuss et al. and Ohtsubo et al. prepared α-Sn films on InSb substrates, and its 3D topological properties were observed.[83,84] In 2015, Zhu et al. used molecular beam epitaxy to successfully epitaxially grow a stanene film on a Bi_2Te_3 (111) substrate, as shown in Figure 3.9a–c, which can be considered the first successful synthesis. The epitaxial monolayer stanene films showed hexagonal atomic structure on the surface, and the lattice constant was the same as that of the substrate. The influence of the substrate on the film lattice constant and electronic structure was then reduced. The stanene film formed on the Bi_2Te_3 substrate may not have the original level of flatness due to the adsorption of H atoms. However, the surface of the film still showed a clear triangular lattice, and no other atomic reconstruction occurred, showing a clear 1×1 stanene structure. The first-principles results inferred that the valence band crossed the Fermi level at the Γ point, indicating the metallic characteristics of the grown stanene.[85] During the experiment, the electronic structure of stanene was affected by the substrate stress, and the topological properties predicted by theory were not obtained, which is a problem to be solved urgently in the follow-up research work. In 2017, Gu et al., reported that the stanene can grow on Sb(111) surface layer by layer.[86] As shown in Figure 3.9d–e, a real honeycomb buckled structure can be observed in the epitaxial structure, indicating that Sn formed a layered stanene structure on the Sb (111) surface.[86] Compared with the growth pattern of Sn on Bi_2Te_3 substrate, the layer-by-layer growth pattern of Sn on Sb (111) effectively solved the previous difficulty of forming a complete Sn monolayer. From the above experimental research, in the growth of 2D materials, lattice adaptation is particularly important. Recently, Liu et al. reported their experimental work on epitaxial growth of stanene on the surface of Au (111) and successfully solved this problem.[87] Sn atoms formed a $22\sqrt{3}$ structure when they deposited to the Au (111) surface at first. When the deposition amount of Sn increased to 0.25 ML, the $22\sqrt{3}$ structure was transformed into a 2×2 alloy structure of Sn and Au. When the Sn deposition amount was further increased

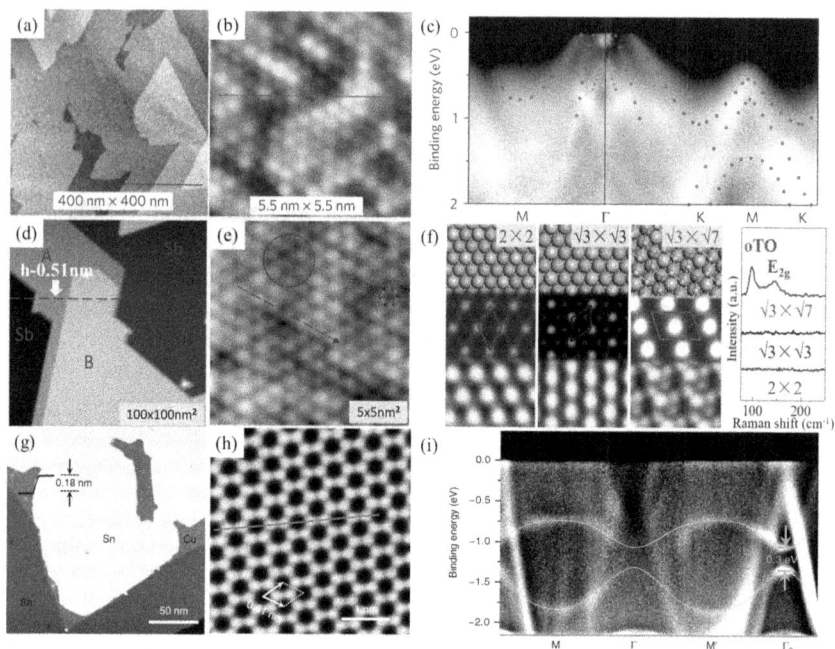

FIGURE 3.9 (a) Large-scale STM topography of stanene film. (b) Atomically resolved STM image of stanene. (c) ARPES spectra along the Γ–M–Γ–K–M–K directions. Black dotted lines mark the experimental electronic bands of stanene. Dashed lines mark one of the hole bands of Bi_2Te_3 (111). Reproduced with permission.[85] Copyright 2015, *Nature*. (d) STM topographic image of 0.5 ML tin grown on Sb (111). (e) High-resolution STM image of stanene. Depression area marked by black circle illustrates a clear honeycomblike atomic structure as illustrated by a ball-and-stick model nearby. Reproduced with permission.[86] Copyright 2015, APS. (f) STM images of stanene epitaxial growth on Au substrate and *in situ* Raman spectra of Au (111) substrate and 2×2, $\sqrt{3}\times\sqrt{3}$, $\sqrt{3}\times\sqrt{7}$, and phases at 532 nm excitation wavelength. Reproduced with permission.[87] Copyright 2019, ACS. (g) STM image of Sn deposited on Cu (111) with 0.9 ML coverage. (h) High-resolution STM image of the stanene film. (i) ARPES of the flat stanene on Cu (111) showing a gap of 0.3 eV at the Γ_2 point. Reproduced with permission.[89] Copyright 2018, *Nature*.

to 0.33 ML, a $\sqrt{3}\times\sqrt{3}$ alloy structure of Sn and Au was formed on the surface of the substrate. Continue to increase the amount of Sn deposited, and the alloy structure changes. When the amount of precipitation exceeds 1 ML, the $\sqrt{3}\times\sqrt{3}$ structure completely disappeared and a new $\sqrt{3}\times\sqrt{7}$ structure appeared. After characterization, it determined that the $\sqrt{3}\times\sqrt{7}$ structure formed at this time was no longer an alloy formed by Sn and Au, but a stanene superlattice with a honeycomb structure, and the 1 × 1 structure of stanene could be seen at the top of $\sqrt{3}\times\sqrt{7}$ superlattice as shown in Figure 3.9f. In addition, it is also a good choice to choose Ag (111) as the substrate for the growth of stanene, which has a lattice constant similar to Au (111) and does not form a superlattice.[88] In 2018, Yuhara et al. confirmed the feasibility of growing stanene on Ag (111). The Ag_2Sn alloy was formed by deposition of 0.33 ML Sn onto Ag (111) substrate at 200°, and the real growth interface of stanene was the Ag_2Sn surface. When 0.5 ML of Sn was deposited on the Ag_2Sn alloy surface at 150°, Sn would grow two-dimensionally along the Ag_2Sn surface terraces. The growth pattern was similar to that of Sn on Sb (111), and no 3D island structure was found. The high-resolution STM images showed that Sn formed an obvious honeycomb structure on the Ag_2Sn surface, but no obvious periodic changes were observed in the height. After measurement, the degree of buckle between the Sn bilayers was less than 0.1 Å, which was much smaller than the previous theoretical prediction of about 0.85 Å, and was closer to the mono-layer film with a planar structure. They also carried out theoretical calculations, and the results were in agreement with the experimental measurements, confirming the reliability of the stanene structure grown on the surface of Ag_2Sn. ARPES measurements showed that there was no momentum scattering in the vertical direction, further demonstrating the 2D structure of the Sn film. However, due to the influence of Ag_2Sn substrate, the Dirac cone structure was not observed on the Sn layer. Although a large-size stanene film can be grown epitaxially on the Ag (111) surface, and the buckling degree of the diatomic layer of Sn films can be greatly reduced, the topological loss of the electronic structure of Sn remained unchanged.

The successful synthesis of stanene materials on different substrates has greatly promoted the development of stanene research, but its topological properties have not been observed in experiments. In 2018, Deng et al. successfully prepared a stanene film with a pure planar honeycomb structure on Cu (111) and observed the topological band inversion in the Cu (111), which confirmed the existence of topological edge states in the material.[89] When Sn was deposited on Cu (111), it showed a typical 2D growth pattern, covering most of the substrate surface as shown in Figure 3.9g–i. In this material system, the stanene was not metallized under the influence of the substrate, and the band gap structure caused by the stanene was observed at the Γ point. This band gap was clearer at Γ_2 point in the second Brillouin region, where a band gap of about 0.3eV can be observed at –1.25eV, which was close to the theoretical prediction of the band gap generated by the introduction of spin-orbit coupling with Sn fluoride. The topological band gap obtained by theoretical simulation is basically in agreement with the experimental results of ARPES, which verifies the topological properties of tin in this system.

Using the experience of graphene, stanene can also be stripped by liquid phase exfoliation. 2D stanene nanosheets were large-scale fabricated by the dealloying of metal lithium in layered Li_5Sn_2 alloy as shown in Figure 3.10.[90] The metal lithium was etched simply via immersion of the layered Li_5Sn_2 alloy into deionized water under ambient atmosphere. After ultrasonication, micron-sized 2D stanene nanosheets with a thickness of 4 nm were successfully prepared. Due to the pseudo-capacitance of the 2D stanene nanosheets, the obtained stanene has high reversible specific capacity and good rate of charge-discharge performance, as well as durable cycling performance.

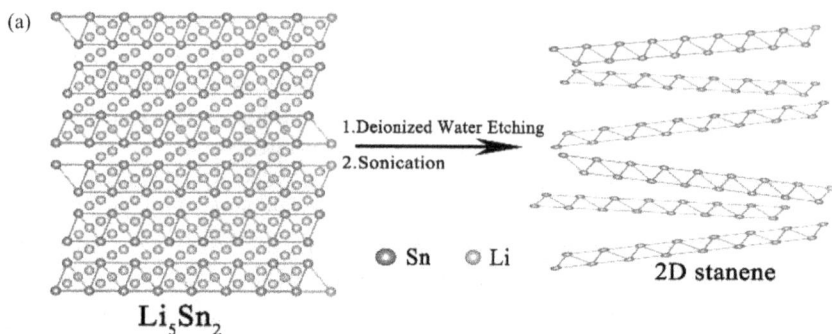

FIGURE 3.10 Fabricating schematic diagram of 2D stanene nanosheets. Reproduced with permission.[90] Copyright 2019, RSC.

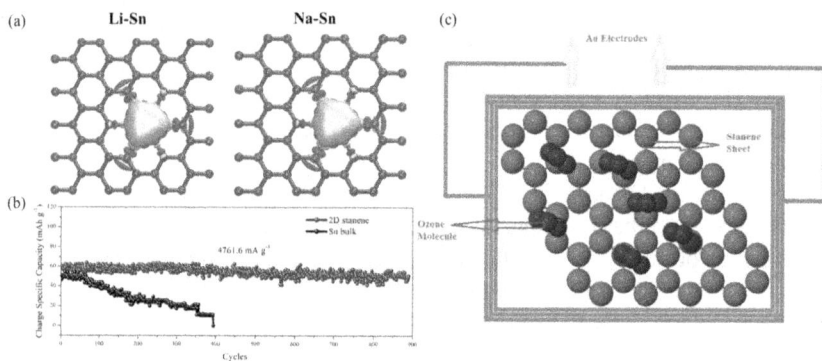

FIGURE 3.11 (a) Calculated charge density difference plots for Li or Na ions on stanene. Reproduced with permission.[91] Copyright 2016, Elsevier. (b) Long cyclic performances of 2D stanene and Sn bulk anode at a current density of 4761 mA g^{-1}. Reproduced with permission.[90] Copyright 2019, RSOC. (c) A schematic illustration of an ozone gas sensor based on stanene nanosheets. Reproduced with permission.[92] Copyright 2019, Elsevier.

3.4.4 APPLICATON OF STANENE

Based on its unique 2D structure, stanene has a variety of exotic properties, excellent thermoelectric properties, topological superconductivity, and near room temperature QAH effect.

3.4.4.1 Batteries

Similar with other 2D materials, stanene are also considered as a promising material for rechargeable energy storage. Mortazavi et al. used first-principles DFT calculations to study and compare the interactions of Na or Li ions on stanene films as shown in Figure 3.11a. In all cases, the hexagonal hollow position provides the maximum binding energy for the ion. And it is predicted that by increasing the ion concentration to the full coverage of the film, the binding energy will remain almost unchanged. Lithium-ion batteries, which are similar in atomic and electronic structure to Sn, are also expected to provide high capacity. The 2D form of stanene, due to its large surface area, high mechanical flexibility, and high electron mobility, can also be considered as a candidate for secondary batteries.[91]

3.4.4.2 Potassium Storage

The dealloying method is a method for preparing 2D materials developed in recent years. It has attracted much attention due to its simple preparation process, scalability, and high cost-effectiveness. In Figure 3.11b, Ma et al. established a simple and feasible dealloying strategy to prepare low-layer stanene.[90] The reversible redox behavior proves that the potassium/depotassium of 2D stanene has good reversibility. In order to gain insight into the potassium/depotassium behavior of 2D stanene nanosheets, cyclic voltammetry (CV) tests were performed at different scan rates (0.2–5 mV s^{-1}). To further study the capacity and cycle performance of 2D stanene nanosheets, constant current charge and discharge tests were carried out at different current densities from 74.4 mA g^{-1} to 4761 mA g^{-1}. Through preliminary comparison, the electrochemical performance of 2D stanene nanosheets on potassium is better than that of traditional carbon materials and Sn-carbon composite materials. 2D stanene anode displayed a high capacity of 300 mAh g^{-1} at 74.4 mA g^{-1} and reversible capacity of 60 mAh g^{-1} at 4761 mA g^{-1} after 900 cycles. This work can provide new insights into the manufacturing of non-van der Waals materials and pave the way for the application of energy storage.[90]

3.4.4.3 Gas Sensors

Toxic air pollutants have serious harmful effects on public health and the environment. The larger surface area of 2D materials strongly proves that they are suitable for advanced gas sensors. Stanene has excellent physical and mechanical properties and has been extensively studied in the field of nanoelectronic devices. In order to give full play to the gas-sensing properties of these materials, Abbasi and Sardroodi conducted theoretical studies on the adsorption behavior of O_3 and SO_3 molecules on 2D stanene sheets as shown in Figure 3.11c. They have studied the structure of the most stable O_3 and SO_3 molecules adsorbed on stanene, and investigated the adsorption process from the energetics, charge transfer, and electronic structure of the adsorption system. The results show that the adsorption of gas molecules on B-doped stanene is more conducive to energy adsorption than that on undoped stanene, indicating that the B-doped system has better sensing performance. B-doped stanene may be an efficient gas sensor device for SO_3 and O_3 in the environment.[92]

Although a series of theoretical and experimental studies have been carried out and some gratifying results have been obtained, there are still some unsolved problems in the intrinsic physical properties, for example, growth preparation and application of stanene. In the experiment, planar single-layer stanene was successfully prepared, showing the topological graphene-like properties of group IV. This study opens the door to explore more 2D topological properties and device applications. There are still few related studies, and the basic physical properties and related composite structures can be further studied.

3.5 PLUMBENE

Lead (Pb) is a nonradioactive element with the largest atomic number in the fourth main group, and the monolayer of Pb in forms of honeycomb structure is called plumbene. Similar with silicene, germanene, and stanene, plumbene prefers to form a buckled structure instead of graphene-like planar structure. However, unlike the stable low-buckling structure of silicene and germanene, as illustrated in Figure 3.12a, a stable plumbene monolayer structure prefers high-buckling.[71] Nevertheless, the research interest on plumbene is mainly focused on

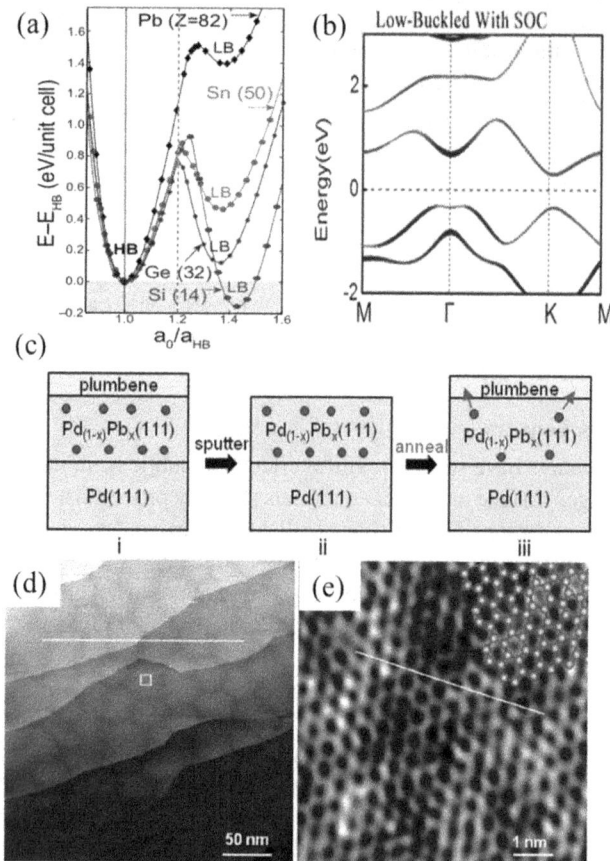

FIGURE 3.12 (a) The high-buckled phase becomes more stable with increasing atomic number. Reproduced with permission.[71] Copyright 2014, APS. (b) Band structure of low-buckled plumbene with spin-orbit coupling. Reproduced with permission.[93] Copyright 2016, *Nature*. (c) Schematic illustration (side view) of: i) plumbene on the $Pd_{1-x}Pb_x$ (111) alloy film on Pd (111), ii) after sputtering to remove the plumbene sheet, and iii) after annealing for plumbene growth by Pb segregation. (d) Large-scale STM image, (e) Atomic-scale STM image. Reproduced with permission.[98] Copyright 2019, Wiley.

modeling or constructing a low-buckling structure because the theoretical results predicted a topological nontrivial energy band in low-buckled plumbene.[93–95]

As shown in Figure 3.12b, the calculated band gap induced by spin-orbit coupling in plumbene is 0.467 eV. Such large value of gaps can be attributed to the strong SOC effect caused by the heavy Pb atoms as well as the px/py orbital character of the low-energy states. The gap is opened at Dirac points and Γ point, which means that plumbene is a two-dimensional topological insulator with Z_2 invariant.[93] Functionalization is an effective way to tune the band structure of plumbene. For example, the band gap of plumbene has been successfully tuned by modifying with F, Cl, Br, I, H, and SiH_3, which can achieve an incasement as high as 1 eV on the band gap size.[93] Wang et al. reported that free-standing plumbene is a topological insulator with a large gap through electron doping, and the nontrivial state is very robust with respect to external strain. Therefore, plumbene was regarded as an ideal candidate for realizing the quantum spin Hall effect at room temperature (RT), promising a great potential for the application of new quantum device.[97]

The first experimentally realized preparation of plumbene was in 2019.[98] Using a palladium single crystal as substrate, Yuhara et al. presented compelling evidence for the successful epitaxial growth of large-area plumbene sheets with a thickness of 0.23 (+0.01) nm. Interestingly, it was found that Weaire–Phelan bubbles' nanostructures are distributed over the terraces of plumbene, as seen in Figure 3.12c–e. According to the STM analysis, these bubble structures are in the range of 15–80 nm in size and about 0.09 nm in height. In addition, plumbene is also obtained on the magnetic Fe single-layer substrate. This is due to the hybridization of the Pb p_z state and the Fe d state, which makes the honeycomb structure stable.[99] So far, the two above-mentioned reports are the only two on the successful experimental realization of plumbene, and more attempts are demanded to promote the investigation on plumbene.

REFERENCES

1. Novoselov, K. S. et al., Two-dimensional gas of massless dirac fermions in graphene. *Nature*, 438, 7065, 2005.
2. Geim, A. K. J. S., Graphene: status and prospects. *Science*, 324, 5934, 2009.
3. Geim, A. K. & Novoselov, K. S., The rise of graphene. *Nanoscience and Technology: A Collection of Reviews from Nature Journals*, 6, 11–19, 2010.
4. Li, D. & Kaner, R. B. J. S., Graphene-based materials. *Science*, 320, 5880, 2008.
5. Kong, X., Liu, Q., Zhang, C., Peng, Z. & Chen, Q., Elemental two-dimensional nanosheets beyond graphene. *Chemical Society Reviews*, 46, 8, 2017.
6. Takeda, K. & Shiraishi, K., Theoretical possibility of stage corrugation in Si and Ge analogs of graphite. *Physical Review B: Condensed Matter*, 50, 20, 1994.
7. Guzmán-Verri, G. G. & Lew Yan Voon, L. C., Electronic structure of silicon-based nanostructures. *Physical Review B*, 76, 7, 2007.
8. Cahangirov, S., Topsakal, M., Akturk, E., Sahin, H. & Ciraci, S., Two- and one-dimensional honeycomb structures of silicon and germanium. *Physical Review Letters*, 102, 23, 2009.

9. Bocchetti, V., Diep, H. T., Enriquez, H., Oughaddou, H. & Kara, A., Thermal stability of standalone silicene sheet. *Journal of Physics: Conference Series*, 491, 2014.

10. Dimoulas, A., Silicene and germanene: silicon and germanium in the "flatland". *Microelectronic Engineering*, 131, 2015.

11. Miro, P., Audiffred, M. & Heine, T., An atlas of two-dimensional materials. *Chemical Society Reviews*, 43, 18, 2014.

12. Balendhran, S., Walia, S., Nili, H., Sriram, S. & Bhaskaran, M., Elemental analogues of graphene: silicene, germanene, stanene, and phosphorene. *Small*, 11, 6, 2015.

13. Ding, Y. & Wang, Y., Density functional theory study of the silicene-like SiX and XSi$_3$ (X = B, C, N, Al, P) honeycomb lattices: the various buckled structures and versatile electronic properties. *The Journal of Physical Chemistry C*, 117, 35, 2013.

14. Zhang, L. et al., Recent advances in hybridization, doping, and functionalization of 2D xenes. *Advanced Functional Materials*, 31, 1, 2020.

15. Drummond, N. D., Zólyomi, V. & Fal'ko, V. I., Electrically tunable band gap in silicene. *Physical Review B*, 85, 7, 2012.

16. Chen, L. et al., Evidence for dirac fermions in a honeycomb lattice based on silicon. *Physical Review Letters*, 109, 5, 2012.

17. Liu, C.-C., Jiang, H. & Yao, Y., Low-energy effective Hamiltonian involving spin-orbit coupling in silicene and two-dimensional germanium and tin. *Physical Review B*, 84, 19, 2011.

18. De Padova, P. et al., Evidence of Dirac fermions in multilayer silicene. *Applied Physics Letters*, 102, 16, 2013.

19. Yi Du, J. Z. et al., Quasi-freestanding epitaxial silicene on Ag(111) by oxygen intercalation. *Research Article*, 2, 7, 2016.

20. Liu, C. C., Feng, W. & Yao, Y., Quantum spin Hall effect in silicene and two-dimensional germanium. *Physical Review Letters*, 107, 7, 2011.

21. Ni, Z. et al., Tunable bandgap in silicene and germanene. *Nano Letters*, 12, 1, 2012.

22. Feng, B. et al., Evidence of silicene in honeycomb structures of silicon on Ag(111). *Nano Letters*, 12, 7, 2012.

23. Lin, C.-L. et al., Structure of silicene grown on Ag(111). *Applied Physics Express*, 5, 4, 2012.

24. Rachid Tchalala, M. et al., Formation of one-dimensional self-assembled silicon nanoribbons on Au(110)-(2 × 1). *Applied Physics Letters*, 102, 8, 2013.

25. Meng, L. et al., Buckled silicene formation on Ir(111). *Nano Letters*, 13, 2, 2013.

26. Li Huang, Y. Z. et al., Sequence of silicon monolayer structures grown on a Ru surface: from a herringbone structure to silicene. *Nano Letters*, 2, 17, 2017.

27. Antoine Fleurence, R. F. et al., Experimental evidence for epitaxial silicene on diboride thin films. *Physical Review Letters*, 108, 2012.

28. Fleurence, A. et al., Microscopic origin of the π states in epitaxial silicene. *Applied Physics Letters*, 104, 2, 2014.

29. Daniele Chiappe, E. S. et al., Two-dimensional si nanosheets with local hexagonal structure on a MoS$_2$ surface. *Advanced Materials*, 26, 2014.

30. Li, W. et al., Ordered chlorinated monolayer silicene structures. *Physical Review B*, 93, 15, 2016.

31. Liu, J., Yang, Y., Lyu, P., Nachtigall, P. & Xu, Y., Few-layer silicene nanosheets with superior lithium-storage properties. *Advanced Materials*, 30, 26, 2018.

32. Zhang, W., Sun, L., Nsanzimana, J. M. V. & Wang, X., Lithiation/delithiation synthesis of few layer silicene nanosheets for rechargeable Li-O2 batteries. *Advanced Materials*, 30, 15, 2018.

33. Su, X. et al., Silicon-based nanomaterials for lithium-ion batteries: a review. *Advanced Energy Materials*, 4, 1, 2014.

34. Tritsaris, G. A., Kaxiras, E., Meng, S. & Wang, E., Adsorption and diffusion of lithium on layered silicon for Li-ion storage. *Nano Letters*, 13, 5, 2013.
35. Liu, J., Yang, Y., Lyu, P., Nachtigall, P. & Xu, Y., Few-layer silicene nanosheets with superior lithium-storage properties. *Advanced Materials*, 30, 26, 2018.
36. Loaiza, L. C., Monconduit, L. & Seznec, V., Siloxene: a potential layered silicon intercalation anode for Na, Li and K ion batteries. *Journal of Power Sources*, 417, 2019.
37. Nematzadeh, M., Massoudi, A. & Nangir, M., Optimization of silicene oxidation as lithium-ion battery anode. *Materials Today: Proceedings*, 42, 2021.
38. Tsai, H.-S. & Liang, J.-H., Production and potential applications of elemental two-dimensional materials beyond graphene. *ChemNanoMat*, 3, 9, 2017.
39. Tao, L. et al., Silicene field-effect transistors operating at room temperature. *Nature Nanotechnology*, 10, 3, 2015.
40. Khan, K. et al., Sensing applications of atomically thin group IV carbon siblings xenes: progress, challenges, and prospects. *Advanced Functional Materials*, 31, 3, 2020.
41. Jose, D. & Datta, A., Structures and electronic properties of silicene clusters: a promising material for FET and hydrogen storage. *Physical Chemistry Chemical Physics*, 13, 16, 2011.
42. Qiu, J. et al., Ordered and reversible hydrogenation of silicene. *Physical Review Letters*, 114, 12, 2015.
43. Kaloni, T. P. & Schwingenschlögl, U., Stability of germanene under tensile strain. *Chemical Physics Letters*, 583, 2013.
44. Ma, Y., Dai, Y., Niu, C. & Huang, B. J. J. O. M. C., Halogenated two-dimensional germanium: candidate materials for being of quantum spin hall state. *Journal of Materials Chemistry*, 22, 25, 2012.
45. Huang, S.-M., Lee, S.-T. & Mou, C.-Y. J. P. R. B., Ferromagnetism and quantum anomalous Hall effect in one-side-saturated buckled honeycomb lattices. *Physical Review B*, 89, 19, 2014.
46. Kaloni, T. P. J. T. J. O. P. C. C., Tuning the structural, electronic, and magnetic properties of germanene by the adsorption of 3d transition metal atoms. *The Journal of Physical Chemistry C*, 118, 43, 2014.
47. Dávila, M. E., Xian, L., Cahangirov, S., Rubio, A. & Le Lay, G., Germanene: a novel two-dimensional germanium allotrope akin to graphene and silicene. *New Journal of Physics*, 16, 9, 2014.
48. Lin, C.-H. et al., Single-layer dual germanene phases on Ag(111). *Physical Review Materials*, 2, 2, 2018.
49. Li, L. et al., Buckled germanene formation on Pt(111). *Advanced Materials*, 26, 28, 2014.
50. Derivaz, M. et al., Continuous germanene layer on Al(111). *Nano Letters*, 15, 4, 2015.
51. Gou, J. et al., Strained monolayer germanene with 1× 1 lattice on Sb (111). *2D Mater*, 3, 4, 2016.
52. Zhuang, J. et al., Dirac signature in germanene on semiconducting substrate. *Advanced Science*, 5, 7, 2018.
53. Yuhara, J. et al., Germanene epitaxial growth by segregation through Ag(111) thin films on Ge(111). *ACS Nano*, 12, 11, 2018.
54. Bampoulis, P. et al., Germanene termination of Ge_2Pt crystals on Ge(110). *Journal of Physics: Condensed Matter*, 26, 44, 2014.
55. Davila, M. E. & Le Lay, G., Few layer epitaxial germanene: a novel two-dimensional Dirac material. *Scientific Reports*, 6, 2016.

56. Qin, Z. et al., Direct evidence of dirac signature in bilayer germanene islands on Cu(111). *Advanced Materials*, 29, 13, 2017.

57. Persichetti, L. et al., van der Waals heteroepitaxy of germanene islands on graphite. *The Journal of Physical Chemistry Letters*, 7, 16, 2016.

58. Zhang, L. et al., Structural and electronic properties of germanene on MoS_2. *Physical Review Letters*, 116, 25, 2016.

59. Bianco, Elisabeth, et al., Stability and exfoliation of germanane: a germanium graphane analogue. *Acs Nano*, 7, 5, 2013.

60. Liu, N. et al., Germanene nanosheets: achieving superior sodium-ion storage via pseudo-intercalation reactions. *Small Structures*, 2, 2100041, 2021.

61. O'Hare, A., Kusmartsev, F. & Kugel, K. J. N. L., A stable "flat" form of two-dimensional crystals: could graphene, silicene, germanene Be minigap semiconductors? *Nano Letters*, 12, 2, 2012.

62. Serino, A. C. et al., Lithium-ion insertion properties of solution-exfoliated germanane. *ACS Nano*, 11, 8, 2017.

63. Zhu, J., Chroneos, A. & Schwingenschlögl, U. J. N., Silicene/germanene on MgX2 (X= Cl, Br, and I) for Li-ion battery applications. *Nanoscale*, 8, 13, 2016.

64. Lin, Z.-Z. & Chen, X., Transition-metal-decorated germanene as promising catalyst for removing CO contamination in H_2. *Materials & Design*, 107, 2016.

65. Jiang, S., Bianco, E. & Goldberger, J. E., The structure and amorphization of germanane. *Journal of Materials Chemistry C*, 2, 17, 2014.

66. Jiang, S. et al., Improving the stability and optical properties of germanane via one-step covalent methyl-termination. *Nature Communications*, 5, 2014.

67. Sahoo, S. K. & Wei, K. H., A perspective on recent advances in 2D stanene nanosheets. *Advanced Materials Interfaces*, 6, 18, 2019.

68. Nakamura, Y. et al., Intrinsic charge transport in stanene: roles of bucklings and electron-phonon couplings. *Advanced Electronic Materials*, 3, 11, 2017.

69. Wang, D., Chen, L., Wang, X., Cui, G. & Zhang, P., The effect of substrate and external strain on electronic structures of stanene film. *Physical Chemistry Chemical Physics*, 17, 40, 2015.

70. Peng, B. et al., Low lattice thermal conductivity of stanene. *Scientific Reports*, 6, 20225, 2016.

71. Rivero, P., Yan, J.-A., García-Suárez, V. M., Ferrer, J. & Barraza-Lopez, S., Stability and properties of high-buckled two-dimensional tin and lead. *Physical Review B*, 90, 24, 2014.

72. Jiang, L. et al., A tight binding and [Formula: see text] study of monolayer stanene. *Scientific Reports*, 7, 1, 2017.

73. Broek, B. v. d. et al., Two-dimensional hexagonal tin:ab initio geometry,stability, electronic structure and functionalization. *2D Materials*, 1, 2, 2014.

74. Jomehpour Zaveh, S., Roknabadi, M. R., Morshedloo, T. & Modarresi, M., Electronic and thermal properties of germanene and stanene by first-principles calculations. *Superlattices and Microstructures*, 91, 2016.

75. Kaloni, T. P. & Schwingenschlögl, U., Weak interaction between germanene and GaAs(0001) by H intercalation: a route to exfoliation. *Journal of Applied Physics*, 114, 18, 2013.

76. Xiong, W. et al., Spin-orbit coupling effects on electronic structures in stanene nanoribbons. *Physical Chemistry Chemical Physics*, 18, 9, 2016.

77. Modarresi, M., Kakoee, A., Mogulkoc, Y. & Roknabadi, M. R., Effect of external strain on electronic structure of stanene. *Computational Materials Science*, 101, 2015.

78. Glavin, N. R. et al., Emerging applications of elemental 2D materials.). *Advanced Materials*, 32, 7, 2020.

79. Xing, D.-X. et al., Tunable electronic and magnetic properties in stanene by 3d transition metal atoms absorption. *Superlattices and Microstructures*, 103, 139,2017.
80. Fu, B., Abid, M. & Liu, C.-C. J. N. J. o. P., Systematic study on stanene bulk states and the edge states of its zigzag nanoribbon. *New Journal of Physics*, 19, 10, 2017.
81. Garg, P., Choudhuri, I. & Pathak, B. J. P. C. C. P., Stanene based gas sensors: effect of spin–orbit coupling. *Physical Chemistry Chemical Physics*, 19, 46, 2017.
82. Xu, Y. et al., Large-gap quantum spin Hall insulators in tin films. *Physical Review Letters*, 111, 13, 2013.
83. Barfuss, A. et al., Elemental topological insulator with tunable fermi level: strained α-Sn on InSb (001). *Physical Review Letters*, 111, 15, 2013.
84. Ohtsubo, Y., Le Fevre, P., Bertran, F. & Taleb-Ibrahimi, A. J. P. r. l., Dirac cone with helical spin polarization in ultrathin α-Sn (001) films. *Physical Review Letters*, 111, 21, 2013.
85. Zhu, F. F. et al., Epitaxial growth of two-dimensional stanene. *Nature Materials*, 14, 10, 2015.
86. Gou, J. et al., Strain-induced band engineering in monolayer stanene on Sb(111). *Physical Review Materials*, 1, 5, 2017.
87. Liu, Y. et al., Realization of strained stanene by interface engineering. *Journal of Physical Chemistry Letters*, 10, 7, 2019.
88. Yuhara, J. et al., Large area planar stanene epitaxially grown on Ag(1 1 1). *2D Materials*, 5, 2, 2018.
89. Deng, J. et al., Epitaxial growth of ultraflat stanene with topological band inversion. *Nature Materials*, 17, 12, 2018.
90. Ma, J., Gu, J., Li, B. & Yang, S., Facile fabrication of 2D stanene nanosheets via a dealloying strategy for potassium storage. *Chem Commun*, 55, 27, 2019.
91. Mortazavi, B., Dianat, A., Cuniberti, G. & Rabczuk, T., Application of silicene, germanene and stanene for Na or Li ion storage: a theoretical investigation. *Electrochimica Acta*, 213, 2016.
92. Abbasi, A. & Sardroodi, J. J., The adsorption of sulfur trioxide and ozone molecules on stanene nanosheets investigated by DFT: applications to gas sensor devices. *Physica E: Low-dimensional Systems and Nanostructures*, 108, 2019.
93. Lu, Y., Zhou, D., Wang, T., Yang, S. A. & Jiang, J. J. S. r., Topological properties of atomic lead film with honeycomb structure. *Scientific Reports*, 6, 1, 2016.
94. Li, Y., Zhang, J., Zhao, B., Xue, Y. & Yang, Z. J. P. R. B., Constructive coupling effect of topological states and topological phase transitions in plumbene. *Physical Review B*, 99, 19, 2019.
95. Zhang, B. et al., The sensitive tunability of superconducting critical temperature in high-buckled plumbene by shifting Fermi level. *Physica E: Low-dimensional Systems and Nanostructures*, 130, 2021.
97. Wang, J., Xu, Y. & Zhang, S.-C. J. P. R. B., Two-dimensional time-reversal-invariant topological superconductivity in a doped quantum spin-Hall insulator. *Physical Review B*, 90, 5, 2014.
98. Yuhara, J., He, B., Matsunami, N., Nakatake, M. & Le Lay, G. J. A. M., Graphene's latest cousin: plumbene epitaxial growth on a "nano watercube". *Advanced Materials*, 31, 27, 2019.
99. Bihlmayer, G. et al., Plumbene on a magnetic substrate: a combined scanning tunneling microscopy and density functional theory study. *Physical Review Letters*, 124, 12, 2020.

4 Group VA of 2D Xenes Materials (Phosphorene, Arsenene, Antimonene, Bismuthene)

Hui Qiao, Rong Hu, Huating Liu,
Xiang Qi, and Jianxin Zhong

CONTENTS

4.1 INTRODUCTION

The interest in monoelement 2D materials has once again injected new vitality into the field of 2D materials, especially the close relative of graphene, group VA 2D Xenes represented by black phosphorus (BP).[1–3] Compared with other monoelement 2D materials, group VA 2D Xenes have richer electronic, optical, and thermodynamic properties. They have a more typical 2D layered structure,

DOI: 10.1201/9781003207122-4

making them easy to prepare on a large scale. The wide range of band gaps that can be adjusted with the number of layers make them not only have typical semiconductor characteristics, but they can also easily achieve the transition from semiconductor to metal phase.[4-6] In particular, the anisotropy brought about by the unique atomic structure has greatly expanded their electronic, optical, and thermodynamic properties and practical applications.

Specifically, black phosphorus has an adjustable direct band gap with the number of layers, and its band gap range is about 0.3–1.8 eV.[5] This is different from traditional 2D TMDs materials (about 1.2–2.5 eV),[7] which can achieve absorption and response from ultraviolet to infrared light. In addition, the high hole mobility of 2D BP exceeds that of most semiconductor 2D materials.[8-11] Subsequently, the same family elements of black phosphorus, 2D As, Sb, and Bi (arsenene, antimonene, bismuthene) were also successfully synthesized and became an important new member of the 2D family. As the number of layers decreases, they present an electronic structure that changes from a metal phase to a semiconductor phase, which indicates that they have rich properties and huge application potential in the field of optoelectronics.[12, 13] Not only that, antimonene and bismuthene also exhibit unique topological properties.[14-16] More interestingly, group VA 2D Xenes have obvious anisotropy in different directions due to their unique atomic arrangement structure,[17-22] which makes them have good application prospects in the fields of polarization optics, electron transport, etc. A surprising discovery, group VA 2D Xenes material shows unexpected advantages in the biomedical field due to its good light-to -heat conversion efficiency. At the same time, the huge specific surface area and biocompatibility of two-dimensional materials can serve as a good platform for nanoparticle drug delivery.[23-25]

The development history of group VA 2D Xenes materials can be traced back to the last century. As early as 1914, Bridgman first synthesized layered bulk black phosphorus with high pressure and high temperature methods.[26] Subsequently, there have been reports constantly predicting the structure and performance of group VA 2D Xenes. Until the successful preparation of graphene, this provided theoretical and experimental guidance for the preparation of group VA 2D Xenes material. At present, the group VA 2D Xenes material field has become one of the fastest growing and most popular research fields and has a wide range of applications in the fields of electronics, optics, energy storage, and biomedicine. Although group VA 2D Xenes materials still have some challenging problems, such as the environmental stability of 2D BP, this still cannot conceal their unique advantages in the field of cutting-edge technology. Moreover, some research reports have been devoted to solving the stability problem of group VA 2D Xenes, and good progress has been made. Here, we review the research progress of group VA 2D Xenes so far, and summarize their crystal structure characteristics, electronic and optical properties, and preparation methods. In addition, the application of group VA 2D Xenes in electronics, optics, energy storage, biomedicine, etc. has been reviewed and summarized, and their superior mechanism analyzed.

4.2 STRUCTURES AND FUNDAMENTAL PROPERTIES OF GROUP VA 2D XENES MATERIALS

4.2.1 THE CRYSTAL STRUCTURE

Group VA 2D Xenes materials, as a branch of 2D materials, have a two-dimensional monoatomic crystal structure similar to grapheme. Slightly different from the planar hexagonal honeycomb lattice composed of sp² hybrid orbitals of graphene carbon atoms, the crystal structure of group VA 2D Xenes material exhibits periodic folding and undulations, with "troughs" extending along the y-axis (b direction).[8] This is because the group VA elements have an extra outer electron, which increases the degeneracy of the electronic state, and the crystal lattice needs to eliminate the degeneracy through periodic fluctuations to make the structure more stable.[27] As we all know, group VA 2D Xenes material has a variety of allotropes, in which the α-phase and β-phase are considered to be stable structures. According to reports, group VA element single-layer material with five typical honeycomb (α, β, γ, δ, ε) and four non-honeycomb (ζ, η, θ, ι) structures have been systematically predicted through theoretical calculations.[28] It is noteworthy that phosphorus has white, red, black, and other amorphous forms, of which BP is the most thermodynamically stable configuration under ambient conditions. Under normal conditions, BP presents a unique puckered structure (α phase), which is converted to β phase under approximately 5 Gpa. Arsenic has three common allotropes, which are gray, yellow, and black arsenic, of which gray arsenic is the most common and stable configuration. For antimony, there are also three well-known allotropes, gray, black and explosive antimony; among them, gray antimony is the most stable one. In the case of bismuth, there is only one stable form that exists, which has the same buckled structure as gray arsenic and gray, namely β phase.

4.2.2 FUNDAMENTAL PROPERTIES

4.2.2.1 Electronic Properties

Band structure and energy level distribution are usually an important parameter of reaction electronic structure. The structure prediction of group VA 2D Xenes materials and first-principles calculations of electronic properties have been widely studied and reported. Zhang et al.[28] found that the band gaps of α-phosphene, α-arsenene, α-bismuthene, and α-antimonene are 1.83, 1.66, 1.18, and 0.36 eV through first-principles calculations at the theoretical level of HSE06 (see Figure 4.1(b)). And the band gaps of β-phosphene, β-arsenene, β-antimonene, and β-bismuthene are 2.62, 2.49, 2.28, and 0.99, respectively. Among them, α-phosphone, α-bismuthene, and β-bismuthene are direct band gaps, and α-arsenene and α-antimonene are quasi-direct band gaps at the Γ point. In addition, the band gap of group VA 2D Xenes materials is closely related to the number of layers. According to reports, BP exhibits a wide band gap ranging from 2 eV to 0.3 eV (Figure 4.1(d)) as the number of layers increases, and it maintains the characteristics of direct band gap during this process,[6] as shown in Figure 4.1(a).

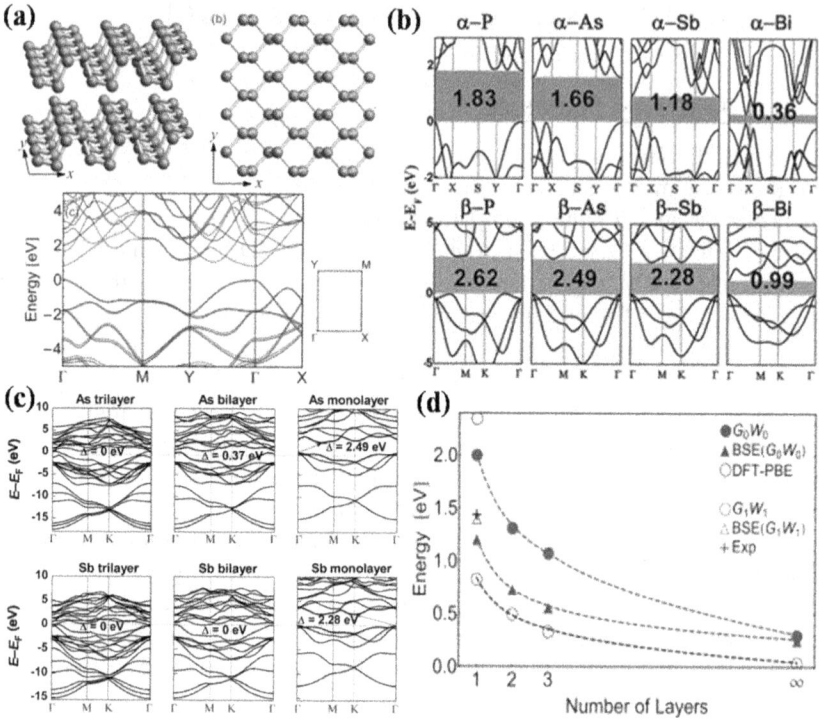

FIGURE 4.1   (a) The bat model with a few layers of phosphorus and the DFT calculation band structure of a single layer of black phosphorus. Reproduced with permission,[6]
copyright 2014, American Physical Society. (b) Calculated band structure of VA 2D Xenes
material in α and β phases. Reproduced with permission,[28] copyright 2016, John Wiley
& Sons. (c) Band gap values of arsenene, antimonene, and bismuthene of different thickness. Reproduced with permission,[30] copyright 2016, John Wiley & Sons. (d) Band gap
values of different layers of black phosphorus obtained by different calculation methods.
Reproduced with permission,[6] copyright 2014, American Physical Society.

This is calculated based on the more accurate GW method, which is consistent
with the experimental results obtained from the scanning tunneling microscope.
Similarly, single-layer arsene, antimonene, and bismuthene all exhibit semiconductor properties. More interestingly, the band gaps of arsene, antimonene, and
bismuthene gradually decrease with the increase of the number of layers, and
when the number of layers is greater than three (Sb is greater than two layers,
as shown in Figure 4.1(c)), they change from a semiconductor phase to a metal
phase.[12, 29, 30] This phenomenon may be due to the increase in the number of layers of arsenene, the enhancement of the interaction force that makes the band gap
gradually decrease, and the quantum confinement effect that causes the valence
band to move to the vacuum level as the thickness decreases.

Among the semiconductor two-dimensional materials, group VA 2D Xenes
materials exhibit excellent carrier mobility. First-principles calculations show that

black phosphorus has excellent hole mobility, up to 26,000 cm^2 V^{-1} s^{-1}[18]. In addition, group VA 2D Xenes materials have obvious anisotropy, which is particularly prominent in terms of carrier mobility. The hole mobility along y is higher (10,000–26,000 cm^2 V^{-1} s^{-1}), which is approximately 16–38 times smaller than the hole mobility along x (640–700 cm^2 V^{-1} s^{-1}). The electron mobility along x is approximately 14 times that of y, which is 1,100–1,140 cm^2 V^{-1} s^{-1} and ~80 cm^2 V^{-1} s^{-1}. In addition, experiments have also confirmed that BP has excellent carrier mobility. Li et al.[31] fabricated a field-effect transistor based on several layers of black phosphorous crystals and found that the carrier can reach up to cm^2 V^{-1} s^{-1}. In addition to BP, other group VA 2D Xenes materials have also been reported to have good carrier mobility, among which the carrier mobility of bismuth is as high as several thousand cm^2 V^{-1} s^{-1}. The calculation results of Pizzi et al.[32] showed that the electron and hole mobilities of β-arsenic were 635 and 1,700 cm^2 V^{-1} s^{-1}, respectively, and β-antimony was 630 and 1737 cm^2 V^{-1} s^{-1}.

4.2.2.2   Optical Properties

The unique optical properties of group VA 2D Xenes materials fully expand their application in the field of optical devices. The light absorption and emission properties of semiconductor materials depend on their optical properties and are also closely related to the electronic energy band structure. Group VA 2D Xenes materials have a wide range of adjustable band gaps with changes in the number of layers so that they have wide-wavelength absorption characteristics, which can achieve absorption and response from ultraviolet to infrared light. However, the light absorption characteristics are affected by the strength of the screening force, size, and exciton effect. In order to correctly describe the optical properties of BP, many interactions (for example, electron and electron-hole interaction) must be considered to obtain accurate light absorption spectra. Çakır et al.[33] found through first-principles calculations that the optical band gap of a single-layer BP is approximately 1.61 eV, which is almost the same as the optical band gap of a single-layer BP obtained experimentally. Low et al.[34] also found that the light absorption capacity of BP is controlled by the thickness. The calculation results found that reducing the thickness of the film is expected to produce a strong exciton effect, which will increase the light absorption near the absorption edge, and it exhibits anisotropic polarized light absorption characteristics as shown in Figure 4.2(a). In addition, Shu et al.[35] confirmed that β-As and β-Sb have broad light absorption properties in light with an energy range of 2–4 eV (including the main part of visible light and ultraviolet light, as shown in Figure 4.2(d)), and β-As and β-Sb excitons. The binding energies are as high as 0.81 eV and 0.73 eV, which can effectively prevent the recombination of photogenerated electron-hole pairs. Xu et al.[36] calculated that the reflectivity of antimony is extremely low, and the absorption in visible light is negligible.

Similar to electronic properties, the light absorption spectrum of group VA 2D Xenes materials shows a strong orientation dependence. Çakır et al.[33] found that due to the strong anisotropy of BP, it has strong absorption of polarized light in the x-direction, and is transparent to polarized light in the y-direction. Qiao et

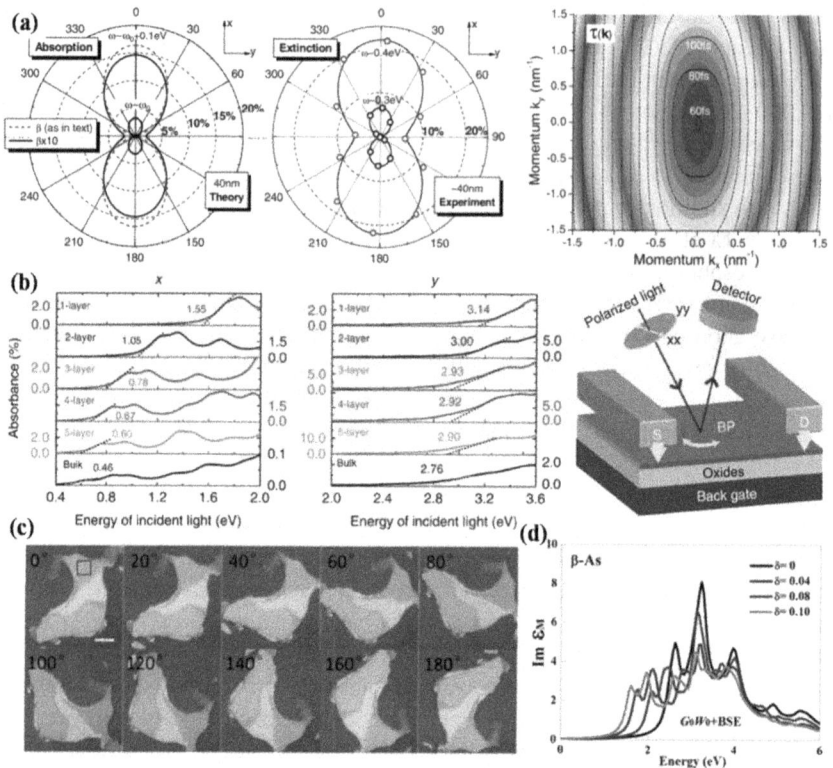

FIGURE 4.2 (a) The polar representation of the 40-nm BP film absorption coefficient A(α) and the polar coordinate representation of the experimental extinction spectrum Z(ω) obtained from the FTIR spectrum, the contour map of the single particle lifetime. Reproduced with permission,[34] copyright 2014, American Physical Society. (b) Optical absorption spectra of few-layer BP for light incident in the c(z)direction and polarized along the a(x)and b(y)directions, respectively. Schematic illustration of a proposed experimental geometry to determine the orientation of few-layer BP structures using optical absorption spectroscopy. Reproduced with permission,[8] copyright 2014, *Springer Nature*. (c) Optical images of a thick BP sample as a function of rotation angle. Reproduced with permission,[37] copyright 2019, American Chemical Society. (d) Optical absorption spectra of β-As for light polarized along the a direction under biaxial tensile strain. Reproduced with permission,[35] copyright 2016, Royal Society of Chemistry.

al.[8] and Tran et al.[6] used first-principles calculations to predict the light absorption spectrum of BP. The calculation results show that the 2D BP has anisotropy in the absorption spectrum of the armchair and linearly polarized light in the zigzag direction, as shown in Figure 4.2(b). This linear dichroism between in-plane directions allows optical determination of crystal orientation and optical activation of anisotropic transmission characteristics. In addition, Mao et al.[37] obtained the inherent anisotropic complex refractive index of BP in the visible light region (480–650 nm) through experiments using anisotropic optical contrast spectroscopy according to the Fresnel equation for the first time, as shown in Figure 4.2(c). In addition to BP, Zhang et al.[38] found through light absorption studies that allotropes of arsenic and antimony have absorption from visible light to ultraviolet light in the x and y directions and have clear linear dichroism.

4.2.2.3 Thermal Properties

In recent years, group VA 2D Xenes materials have been widely researched and applied in sensors, biomedicine, and other fields due to their good thermodynamic properties. In particular, 2D black phosphorus not only exhibits excellent thermodynamic properties, but also makes its thermodynamic properties colorful due to its outstanding anisotropy. Aierken et al.[39] observed that black phosphorus has good thermodynamic stability and highly anisotropic thermal properties. The difference in the linear thermal expansion coefficient along the zigzag and armchair directions is up to 20%. Zhu et al.[40] studied the thermal conductivity of monolayer black phosphorus by combining the density functional calculation and the Peierls-Boltzmann transport equation. The calculation results show that the single-layer black phosphorus exhibits size-dependent and size-independent thermal conductivity along the zigzag and armchair directions, respectively. For a single layer BP with a size of 10 μm, the thermal conductivity at 300 K along the zigzag and armchair directions are 83.5 and 24.3 W/mK, respectively. Through first-principles calculations, Sun et al.[41] found that the thermal conductivity tensor of bulk black phosphorus has a considerable phonon frequency dependence in the range of 80–300 K. Subsequently, through a series of experimental methods, the thermodynamics of group VA 2D Xenes materials were observed and their thermodynamic properties were comprehensively evaluated. Jang et al.[42] used conventional TDTR and beam shift methods to detect the anisotropic heat transfer performance of mechanically peeled multilayer BP sheets at room temperature. The results show that the highest in-plane thermal conductivity in the zigzag and armchair directions are 86 ± 8 and 34 ± 4 W m^{-1} K^{-1}, respectively, which not only shows a very high in-plane thermal conductivity value, but also has three times the in-plane thermal conductivity. In-plane anisotropy. Luo et al.[43] found that for a black phosphor film with a thickness greater than 15 nm, the thermal conductivity of the armchair and the zigzag shape are 20 and 40, respectively. As the film thickness decreases, the thermal conductivity of the armchair and the zigzag shape decrease to 10 and 20, respectively. And the thermal conductivity anisotropy ratio of the black phosphor film with a thickness greater than 15 nm is about 2, while the thermal conductivity anisotropy ratio of the thinnest 9.5 nm film is reduced to 1.5. Not only that, there are reports that as the temperature increases, the anisotropy of the thermal conductivity of black phosphorus will increase.[44]

In addition, some studies have shown that arsene, antimonene, and bismuthene, which are the same family elements of black phosphorus, also exhibit good thermodynamic properties. For example, studies have shown that arsene has a smaller, single but more anisotropic thermal conductivity, and its thermal conductivity along the zigzag and armchair directions at room temperature is 30.4 and 7.8 W m^{-1} K^{-1}, respectively.[22] At the same time, Wang et al.[45] predicted that antimonene has a lattice thermal conductivity of 15.1 W m^{-1} K^{-1} at 300 K. Similar to arsene and antimonene, bismuthene has a smaller thermal conductivity than black phosphorus, but can easily adjust its thermal conductivity through a series of strategies, such as the n/p-doping.

4.3 PREPARATION OF GROUP VA 2D XENES MATERIAL

Group VA 2D Xenes materials have developed to date and have become an independent and mature system, where the preparation of high-quality group VA 2D Xenes materials have been realized through a variety of methods to meet different application requirements. This chapter will briefly introduce the various preparation processes of group VA 2D Xenes materials and the advantages and disadvantages of these preparation strategies.

4.3.1 TOP-DOWN

The top-down method is usually to separate the thinner few-layer or single-layer 2D Xenes nanosheets from the existing 3D bulk materials. At the earliest, 2D graphene was prepared for the first time by mechanical exfoliation, and the concept of 2D monoatomic layer material was proposed. This mechanical exfoliation method has wide applicability and is also widely used in the preparation of group VA 2D Xenes materials,[2, 31, 46] as shown in Figure 4.3(a) and (b). However, the yield of this method is very low, and large-scale preparation and application cannot be realized. Therefore, a liquid phase exfoliation method with a simple process and large-scale preparation was developed and widely used in the preparation of 2D materials[47] (see Figure 4.3(c) and (d). For example, Brent et al.[48] used N-methyl-2-pyrrolidone (NMP) as a dispersant to peel off black phosphorus, and produced three to five layers of black phosphorus nanosheets with obvious lateral dimensions. At the same time, Hanlon et al.[49] used liquid phase exfoliation in solvents such as N-cyclohexyl-2-pyrrolidone (CHP) to produce large-scale BP nanosheets with controllable and small layers. And it was surprisingly found that the solvated shell protects the nanosheets from reacting with water or oxygen, so that the black phosphorous nanosheets exhibit good stability. Sun et al.[3] combined probe ultrasonic treatment and bath ultrasonic treatment to synthesize ultra-small black phosphorous quantum dots in N-methyl-2-pyrrolidone solution with a lateral size of about 2.6 nm. In addition, the large-scale preparation of arsenene, antimonene, and bismuthene has also been achieved by liquid phase exfoliation. For example, Beladi-Mousavi et al.[50] reported the preparation of high-quality, high-stability, several-layer arsenic nanosheets through a liquid phase stripping procedure (Figure 4.3(f)) with the aid of ultrasonic treatment in an N-methylpyrrolidone solution without any other surfactants. Gibaja et al.[51] produced a highly stable isopropanol/water (4:1) suspension of several layers of antimony through liquid phase exfoliation of antimony crystals with the aid of ultrasonic treatment, as shown in Figure 4.3 (g). Zhang et al.[52] realized the large-scale preparation of large-scale ultra-thin two-dimensional bismuth (Bi) nanosheets through a liquid

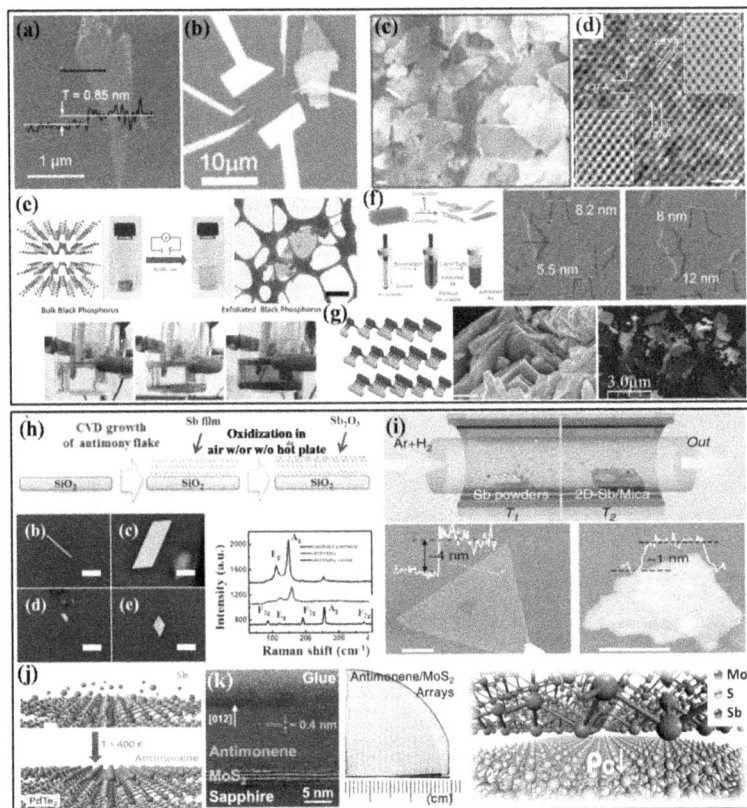

FIGURE 4.3 (a) Atomic force microscopy image of a single-layer phosphorene crystal. Reproduced with permission,[46] copyright 2019, American Chemical Society. (b) Optical image of BP device manufactured by mechanical exfoliation process. Reproduced with permission,[2] copyright 2014, *Springer Nature*. (c) High-magnification Scanning Electron Microscope (SEM) image and (d) high-resolution Transmission Electron Microscopy (TEM) image of BP nanosheets deposited on SiO_2/Si substrate. Reproduced with permission,[47] copyright 2015, John Wiley & Sons. (e) Schematic diagram of electrochemical exfoliation of BP nanosheets. Reproduced with permission,[53] copyright 2017, John Wiley & Sons. (f) Schematic and characterization diagram of 2D-Arsenene prepared by liquid phase exfoliation. Reproduced with permission,[50] copyright 2019, John Wiley & Sons. (g) Atomic structure diagram and characterization image of 2D-Antimonene prepared by liquid phase exfoliation. Reproduced with permission,[51] copyright 2016, John Wiley & Sons. (h) Schematic diagram and characterization image of growth of single crystal antimony sheet by chemical vapor deposition method on SiO_2 substrate. Reproduced with permission[58] copyright 2018, Royal Society of Chemistry. (i) Schematic diagram and AFM characterization image of epitaxially grown 2D antimonene single crystal. Reproduced with permission,[59] copyright 2016, *Springer Nature*. (j) Schematic diagram of epitaxial growth of a single layer of antimonene on $PdTe_2$. Reproduced with permission,[64] copyright 2017, John Wiley & Sons. (k) Schematic diagram and HRTEM image of Antimonene film grown by molecular beam epitaxy. Reproduced with permission,[65] copyright 2018, American Chemical Society.

stripping strategy. Up to now, liquid phase exfoliation has become a mature large-scale preparation method for the preparation of group VA 2D Xenes materials.

Recently, some other top-down preparation methods for group VA 2D Xenes materials have also been developed and reported. For example, the electrochemical exfoliation method.[53] (Figure 4.3(e)). Erande et al.[54] used electrochemical exfoliation to synthesize ultra-thin black phosphorus nanosheets with a small atomic layer thickness. Marzo et al.[55] prepared few-layer Sb nanosheets by anodic-cation

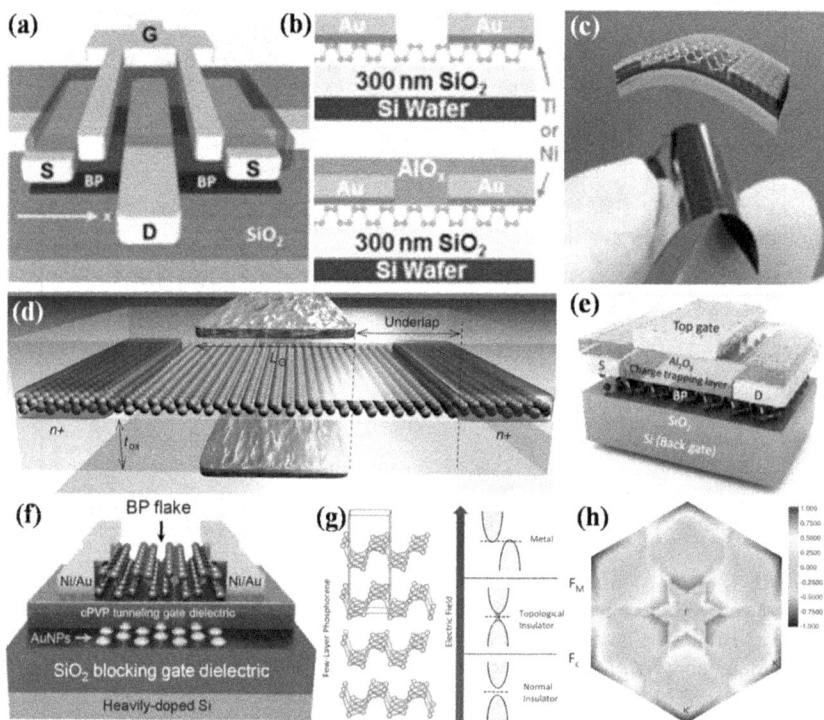

FIGURE 4.4 (a) Schematic of the BP transistor device structure. Reproduced with permission,[73] copyright 2014, American Chemical Society. (b) Schematics of unencapsulated and AlOx encapsulated BP field-effect transistors. Reproduced with permission,[70] copyright 2014, American Chemical Society. (c) Simplified illustration of the device structure of flexible BG BP FET on PI substrate. Reproduced with permission,[74] copyright 2015, American Chemical Society. (d) Schematic view of the double-gate n-doped MOSFETs. Reproduced with permission,[32] copyright 2016, *Springer Nature*. (e) Schematic of the ambipolar BP charge-trap memory device. Reproduced with permission,[76] copyright 2016, American Chemical Society. (f) Schematic structure of the nano floating gate transistor memories comprising few-layer BP flakes. Reproduced with permission[77] copyright 2016, Royal Society of Chemistry. (g) Crystal structure diagram of phosphorene multilayer stack. Reproduced with permission,[78] copyright 2015, American Chemical Society. (h) Degree of circular polarization η(k) in irreducible Brillouin zone of BiH monolayer under a vertical electric field. Reproduced with permission,[81] copyright 2014, *Springer Nature*.

intercalation exfoliation in an Na_2SO_4 electrolyte. Subsequently, Kovalska et al.[56] realized the electrochemical exfoliation of bulk black arsenic in a non-aqueous electrolyte medium for the first time, and prepared a stable dispersion of several layers of arsenene.

4.3.2 BOTTOM-UP

The bottom-up method of preparing 2D materials can usually be described as a "grow out of nothing" process. The bottom-up strategy is to grow from the atomic or molecular level to obtain 2D nanosheets. For example, Li and Smith et al. used in-situ chemical vapor deposition to grow a large area of two-dimensional black phosphorus film[2,57], Wu[58] directly grew antimony nanosheets on a SiO_2 substrate by chemical vapor deposition (Figure 4.3 (h)), in which the antimony nanosheets was stacked to a thickness of several tens of nanometers, with a typical area of about 40 μm. In addition, Ji et al.[59] synthesized high-quality layers of antimonene on a mica substrate through van der Waals epitaxy, as shown in Figure 4.3(i). And through HRTEM microscope and Raman spectroscopy analysis, it is shown that the obtained antimonene polygon has a curved hexagonal structure (β phase), which is consistent with the atomic structure of the most stable single-layer antimonene allotrope predicted by previous theoretical studies. In addition, our team prepared a single-crystal, high-quality, and continuous 2D antimony (Sb) film on SiO_2 and flexible polydimethylsiloxane substrates by vapor deposition.[60] Subsequently, we synthesized the Bi film again successfully on SiO_2/Si (SiO_2) and polyimide (PI) substrates through a direct and simple vapor deposition method.[61] Recently, Kuriakos et al.[62] used a physical vapor deposition process to achieve the controlled growth of millimeter-level β-phase single crystal antimonene nanosheets on a SiO_2 dielectric substrate. In addition, there are many reports confirming that the preparation of group VA 2D Xenes materials by vapor deposition method is an efficient and mature strategy.

Although the vapor deposition method is already a very mature method to prepare group VA 2D Xenes materials, it also has great limitations. The experimental conditions of the vapor deposition method are usually carried out under high temperature conditions, and it is difficult to achieve precise controllable adjustment for the growth of 2D nanomaterials. In contrast, a molecular beam epitaxial growth method overcomes these shortcomings very well, and it can grow under low temperature conditions, which greatly increases the selectivity of the substrate material. At the same time, the film growth rate of molecular beam epitaxial growth is slow, the beam intensity is easy to accurately control, and the film composition and doping concentration can be adjusted quickly with the change of the molecular source. Yang et al.[63] used black phosphorus as the precursor to grow a single layer of blue phosphorus with the same layered structure and high stability as black phosphorus on Au (111) by molecular beam epitaxy. In addition, Wu et al.[64] reported that a single layer of antimonene was grown on a 2D layered $PdTe_2$ substrate by molecular beam epitaxy, as shown in Figure 4.3(j). Subsequently, Chen et al.[65] prepared single crystal antimonene on a sapphire substrate by molecular beam epitaxy. Recently, Zhong et al.[66] successfully grown 2D bismuthene with a thickness of ≤30 nm on Si(111) by extra-molecular beam method, as shown in Figure 4.3(k).

4.4   APPLICATIONS OF GROUP VA 2D XENES MATERIALS

Group VA 2D Xenes materials have potential applications in many fields due to their excellent electronic, optical, and thermodynamic properties. At the same time, due to the anisotropy caused by their unique structure, their properties become more and more colorful. In recent years, group VA 2D Xenes materials have been widely used in electronics, optoelectronics, energy, and biomedical fields.

4.4.1   ELECTRONIC DEVICES

As a typical electronic device, the invention of the field effect transistor has far-reaching significance to the development of the electronic industry, and is widely used in the core components of various electronic products. Li et al.[31] used adhesive tape to peel off the black phosphorous flakes from the bulk single crystal and fabricated a field-effect transistor based on several layers of black phosphorous crystals. The device exhibits bipolar behavior, with a drain current modulation as high as 10^5, and it is found that the highest carrier mobility can reach ~1,000 cm^2 V^{-1} s^{-1} at room temperature when the thickness is 10 nm. Du et al.[67] found that the switching mechanism of the BP transistor is controlled by two Schottky barriers on the metal contacts. By appropriately modulating the channel length, gate bias, and drain bias, the device characteristics can be changed from the inherent p-type The behavior changes to n-type behavior. Perello et al.[68] also reported a unipolar n-type black phosphorous transistor that achieves switch polarity control through contact metal engineering and sheet thickness. However, black phosphorus is prone to degradation under the action of H_2O and O_2, which seriously hinders its further development. Therefore, some measures must be taken to inhibit and reduce its surface degradation. For example, Na et al.[69] and Wood et al.[70] confirmed through the same Raman spectroscopy that Al_2O_3 passivated several-layer black phosphorus nanochip-based field-effect transistors have a shelf life of up to two months in ambient air (Figure 4.4(b)). It is worth noting that He et al.[71] deposited an organic thin film (dioctylbenzothienobenzothiophene, C8-BTBT) on black phosphorus nanosheets, which can effectively protect black phosphorus from oxidation under environmental conditions for 20 days. At the same time, the non-covalent van der Waals interface between C8-BTBT and black phosphorus effectively retains the inherent characteristics of black phosphorus, which makes the current density reach a record high of 920 μA/um, and the hole drift speed exceeds 1×10^7 cm/s at room temperature, and the on/off ratio is 1×10^4 and ~1×10^7. Not only that, but black phosphorus is also considered a promising material for high-frequency electronic equipment due to its excellent electronic properties.[72] Wang et al.[73] demonstrated for the first time the gigahertz frequency operation of a BP field-effect transistor (Figure 4.4(a)) and achieved a current density of more than 270 mA/mm and a DC transconductance of more than 180 mS/mm for hole conduction. At present, flexible wearable devices are more popular than traditional electronic products due to their lightness and flexibility. Zhu et al.[74] reported the first flexible black phosphorous FET with good electron and hole mobility, and realized the basic circuit and amplitude modem of the electronic system for flexible technology, as shown in Figure 4.4(c). In addition to black phosphorus-based FETs, other FETs based on group VA 2D Xenes materials have also been built. Pizzi et al.[32] fabricated field-effect transistors based on a single layer of arsenic and antimony as the channel, and showed excellent carrier mobility and switching properties (Figure 4.4(d)). Li et al.[75] predicted the possibility of applying arsenene, antimonene, and bismuthene to 10 nm gate-long tunneling field effect transistors by calculating quantum transfer simulations. Among them, the ML bismuth TFET has the largest on-state current ion of about 1153 $\mu A\ \mu m^{-1}$

Group VA 2D Xenes materials also have good applications in other electronic devices. Tian et al.[76] demonstrated a bipolar black phosphorus charge trap storage device with dynamically reconfigurable and polarity reversible storage behavior (Figure 4.4(e)). The memory can continuously adjust the current ratio of the programmed/erased state to achieve interchangeable high and low current levels in the erased and programmed states, and this memory device based on layered materials has a multi-level cell storage capability. Lee et al.[77] fabricated a nano-floating gate transistor storage device using several layers of black phosphorus channels and a gold nanoparticle charge trap layer (Figure 4.4(f)), and demonstrated excellent storage performance, including five levels of data storage, large storage window, stable retention time, and cycle durability. In addition, some theoretical calculations have predicted that arsenic, antimony, and bismuth in group VA 2D Xenes materials have unique topological properties and are promising materials for spintronic devices. For example, the report by Liu et al.[78] confirmed that the phase change from normal to topological can be achieved along the stacking direction of several layers of black phosphors along the electric field (Figure 4.4(g)). Zhang et al.[79] calculated by density functional theory, and we predict that appropriate strain modulation of honeycomb arsenic will result in a unique two-dimensional topological insulator that can be characterized and utilized at room temperature. Zhao et al.[80] found that the bending structure of antimonene enables it to withstand large tensile strains of up to 18%, and the resulting bulk band gap can be as high as 270 meV, thereby realizing a 2D topological phase transition at room temperature. In addition, there are reports that arsenic, antimony, and bismuth are modified into topological insulators through different chemical functionalizations. Song et al.[81] predicted a set of two-dimensional topological insulator BiX/SbX (X = H, F, Cl and Br) monolayers with a body gap ranging from 0.32 eV to a record 1.08 eV (Figure 4.4(h)).

4.4.2 OPTOELECTRONICS DEVICES

Group VA 2D Xenes materials have attracted much attention in the field of optoelectronics due to their unique in-plane anisotropy, thickness-dependent direct band gap, and high carrier mobility. As the most eye-catching optoelectronic device, photodetectors have been widely used in many fields such as laser sensor originals, infrared imaging, missile tracking, etc. For single-layer black phosphorous nanosheets, the band gap is about 1.8 eV, which can be applied to UV-visible light detectors. Wu et al.[82] proved for the first time that black phosphorus can be used in ultraviolet photodetectors (Figure 4.5(a)), and has excellent light response performance, and its specific detection ratio is about 3×10^{13} Jones. Buscema and Low et al.[83, 84] constructed a transistor based on a few layers of black phosphorous, which showed a response to an excitation wavelength from the visible light region

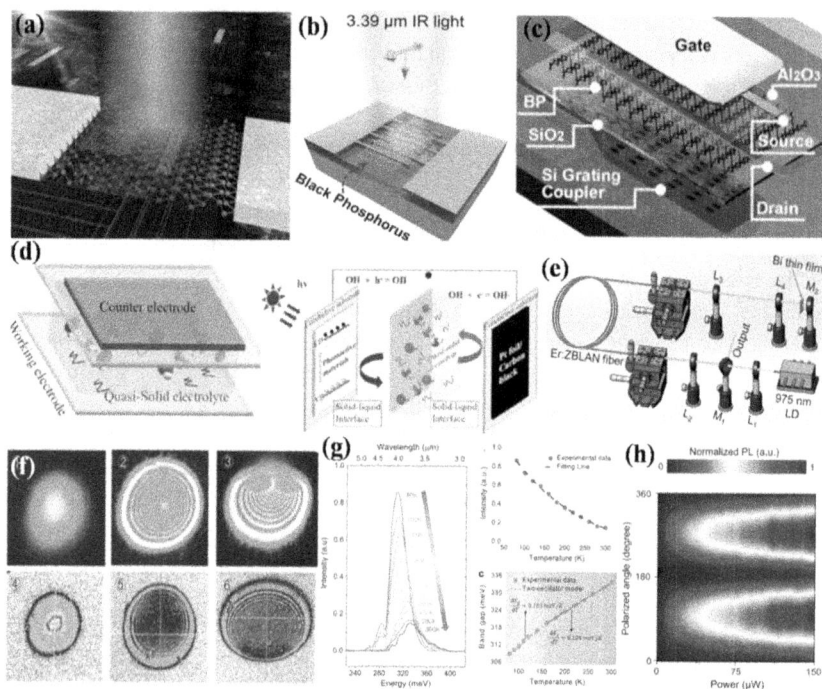

FIGURE 4.5 (a) Three-dimensional view of the structure of the BP UV photodetector. Reproduced with permission,[82] copyright 2015, American Chemical Society. (b) BP Mid-Infrared Photodetectors with High Gain. Reproduced with permission,[86] copyright 2016, American Chemical Society. (c) Waveguide-integrated black phosphorus photodetector for mid-infrared applications. Reproduced with permission,[87] copyright 2018, American Chemical Society. (d) Schematic diagram charge transfer of as-fabricated PEC-type photodetector. Reproduced with permission,[96] copyright 2021, Elsevier B.V. (e) Schematic diagram of fiber laser based on Bi film. Reproduced with permission,[101] copyright 2019, American Chemical Society. (f) SSPM diffraction ring pattern of antimonene dispersions with a continuous wave. Reproduced with permission,[102] copyright 2017, John Wiley & Sons. (g) Temperature dependence of PL spectrum of BP film. Reproduced with permission,[104] copyright 2019, American Chemical Society. (h) Linear polarization-dependent mapping of BP photoluminescence intensity under various excitation powers. Reproduced with permission,[105] copyright 2020, American Chemical Society.

to 940 nm, a rise time of about 1 ms, and a responsivity of 4.8 mA/W under illumination. The band gap of black phosphor with adjustable number of layers makes it have a wide wavelength range of light response performance, and it is also a suitable candidate for infrared photoelectric applications to fill the gap[85–87](Figure 4.5(b) and (c)). Engel et al.[88] studied a multilayer black phosphorous photodetector that can acquire high-contrast images in the visible spectrum (λ_{VIS} = 532 nm) and infrared spectrum (λ_{IR} = 1550 nm). Li et al.[89] studied the performance regulation of black phosphorous photodetectors with the number of layers, and the intrinsic responsivity was as high as 135 mA W^{-1} and 657 mA W^{-1} in 11.5 and 100 nm thick devices, respectively. The unique anisotropic carrier transport and light absorption behaviors of black phosphor are derived from the in-plane asymmetric structure, so phosphor can be used as a broadband polarization-sensitive photodetector from 400 nm to 3750 nm. Chen et al.[90] further studied the regulation mechanism of the electric field on the infrared light response performance of black phosphor. The experimental results found that the vertical electric field can extend the light response in the 5 nm thick black phosphor photodetector from the cut-off wavelength of 3.7 μm to 7.7μm. A series of means have been proposed to improve the stability of BP photodetectors. Na et al.[91] have achieved the air stability of several layers of BP phototransistors through proper Al$_2$O$_3$ passivation. The device has strong robustness in the surrounding air and can be stored for more than six months. Subsequently, Zhan et al.[92] proposed an interesting "Self-Healable" concept to improve the stability of black phosphorous photodetectors. A hydrophobic polyionic liquid poly hexafluorophosphate was applied to encapsulate BP quantum dots for the construction of photoelectrochemical photodetectors, and showed long environmental stability up to 90 days. In addition to black phosphorous, other group VA 2D Xenes materials have also shown good applications in the field of photodetectors. Xiao et al.[93] produced for the first time a flexible photodetector based on a mixed structure of a small amount of antimony nanosheets modified with CdS quantum dots. The detector showed a good response of 10 μA W^{-1} and an on/off ratio of 26.8 under 1 V bias. Recently, a photoelectrochemical (PEC) photodetector based on group VA 2D Xenes materials has been widely reported. First of all, our team produced for the first time a PEC-type photodetector based on a few layers of black phosphorous. This photodetector exhibits good self-powered light response performance and environmental stability for up to one month in KOH solution.[94] Subsequently, we manufactured Bi-based PEC-type photodetectors[95] and flexible PEC-type photodetectors,[96] which also showed good light response performance and expanded to the field of flexible optoelectronics, as shown in Figure 4.5(d). In addition, Su et al.[97] synthesized surface-modified multilayer antimony by electrochemical exfoliation and synchronous halogenation and used it in PEC-type photodetectors. Subsequently, they reconstituted a PEC-type photodetector based on Bi QDs that not only exhibits the appropriate ability to self-drive broadband light response, but also exhibits high-performance light response and long-term stability behavior from UV to visible light.[98]

Studies have shown that group VA 2D Xenes materials have nonlinear optical properties, and it has been observed that their nonlinear optical properties are all anisotropic. Li et al.[99] used the nonlinear optical properties of black phosphorus for ultra-fast and high-energy pulse generation in fiber lasers, and observed that black phosphorus has relatively large optical nonlinearity and unique polarization and thickness dependence. Subsequently, Jiang et al.[100] confirmed that black phosphorous quantum dots have great potential in the application of broadband light limiters in the visible range or ultrafast laser pulses of saturable absorbers in the near-infrared range. Du et al.[101] prepared a highly stable single-crystal continuous Bi film by vapor deposition and found that it can exhibit a broad-band, ultra-fast nonlinear optical response under strong excitation and has a low saturation intensity from the near-infrared to the mid-infrared spectrum (Figure 4.5(e)). Lu et al.[102] found that antimonene has a huge nonlinear refractive index of $\approx 10^{-5} cm^2 W^{-1}$, and it can last for several months under environmental conditions, and is considered to be a promising material for nonlinear photonics equipment (Figure 4.5(f)).

Group VA 2D Xenes materials are semiconductors with a significant base band gap, making them a potential candidate material for light-emitting devices. Zhang et al.[103] observed strong and highly layer-dependent photoluminescence in several layers of black phosphorus (two to five layers). Later, Chen et al.[104] discovered the thickness-dependent photoluminescence spectrum in the film BP under mid-infrared conditions, as shown in Figure 4.5(g). In recent years, light-emitting diodes based on black phosphorus have been widely reported. The van der Waals heterostructure based on black phosphor emits linearly polarized light, and the spectrum covers the technically important mid-infrared atmospheric window. It is used as a light-emitting diode with fast modulation speed and excellent operational stability[105] (Figure 4.5(h)). In addition, Tsai et al.[106] produced multilayer arsenic nanobelts and showed green PL emission at 540 nm at room temperature, indicating that the band gap of the multilayer arsenic nanobelts is about 2.3 eV. Tsai et al.[107] also produced a multilayer antimony nanoribbons that emits orange light at 610 nm. Liu et al.[108] produced fluorescent nanoprobes based on antimonene quantum dots, which showed stable blue fluorescence, and the photoluminescence was hardly affected by pH. Hussain et al.[109] found that ultra-thin Bi nanosheets exhibited morphologically and structure-dependently enhanced wide-range light emission in the visible spectrum.

4.4.3 ENERGY CONVERSION AND STORAGE

Group VA 2D Xenes materials have a higher theoretical capacity than graphite, which indicates that they are promising candidates for battery electrode materials. Through density functional theory calculations, Li et al.[110] found that Li ions have an ultra-high diffusivity along the zigzag direction in the black phosphorus nanosheets, and predicted that black phosphorus will be expected to become a lithium-ion battery with high rate capability and high charging voltage. The calculation results show that the diffusion barrier of Li atoms along the zigzag energy is calculated to be 0.08 eV, which is much lower than the other two lithium battery anode materials, graphene and MoS_2.[111–113] In addition, the intercalation of lithium ions can cause the transformation of black phosphorus from semiconductor to metal, resulting in good electrical conductivity, which is extremely beneficial for lithium ion battery electrodes. Liu et al.[114] confirmed by first-principles calculations that there is a strong binding energy between Na^+ and phosphorus; the Na^+ can be stabilized on the phosphorus surface without agglomeration, and the Na^+ insertion capacity is about 324 mA h g^{-1}. In experiments, Chen et al.[115] combined black phosphorous nanosheets with highly conductive grapheme (Figure 4.6(b)), which proved that black phosphorous nanosheets can be used on high-performance flexible Li ion battery (LIB) electrodes, and exhibited a high specific capacity of 501mA h g^{-1} and excellent magnification. Performance and long-cycle performance at a current density of 5 mA h g^{-1}. Sultana et al.[116] constructed a high-capacity potassium ion battery based on black phosphorus and provided a first cycle capacity of up to 617 mA h g^{-1}. Recently, Callegari et al.[117] used a characteristic organic solvent as a binder to construct a self-healing black phosphorus anode sodium ion battery, as shown in Figure 4.6(c). The battery showed a capacity that was increased by more than six times, and had better adhesion, buffering performance, and spontaneous damage repair ability. Recently, Benzidi et al.[118] and Ye et al.[119] demonstrated through density functional theory calculations that high capacity, low open circuit voltage, and ultra-high barrier diffusion make arsenic a good candidate for use as an anode material for rechargeable batteries. Tian et al.[120] confirmed that 2D antimonene is a promising anode material in sodium-ion batteries because of its theoretical capacity of up to 660 mAh g^{-1} and larger surface active sites (Figure 4.6(a)). In addition, Zhou et al.[21] found that a free-standing bismuth/graphene composite electrode with adjustable thickness can achieve amazing stability and a high-area sodium storage capacity of 12.1 mAh cm^{-2}, which greatly exceeds most reported electrode materials. This is mainly due to the ultra-thin layers of bismuthene with a large aspect ratio and can reduce the expansion strain along the z-axis. Not only that, but there are also many reports that the moderate adsorption energy of arsenene, antimonene, and bismuthene on all polysulfides, which indicates that they can effectively inhibit the shuttle effect and become a promising cathode anchor material for improving the performance of Li-S batteries.[121, 122]

Photocatalytic/electrocatalytic water splitting is the core process of many energy conversion devices. Group VA 2D Xenes materials are used in the field of catalysis due to large specific surface area, good light absorption performance, and chemical activity.[123] Zhu et al.[124] reported for the first time that the black phosphorus nanosheets prepared by the ball milling method have a visible light photocatalytic hydrogen evolution rate of 512 μmol h^{-1} g^{-1} without using any precious metal promoters, which is 18 times that of bulk black phosphorus, and is comparable or even higher than that of g-C$_3$N$_4$. The enhancement of the photocatalytic performance of BP nanosheets due to the negative shift of the conduction band level is more feasible for reducing the reduction potential of H$^+$/H$_2$, and the positive shift of the valence band level can effectively inhibit the recombination of electron-hole pairs. In addition, Wen et al.[125] used black phosphorous quantum dots as hole transfer promoters to effectively promote the separation of photogenerated

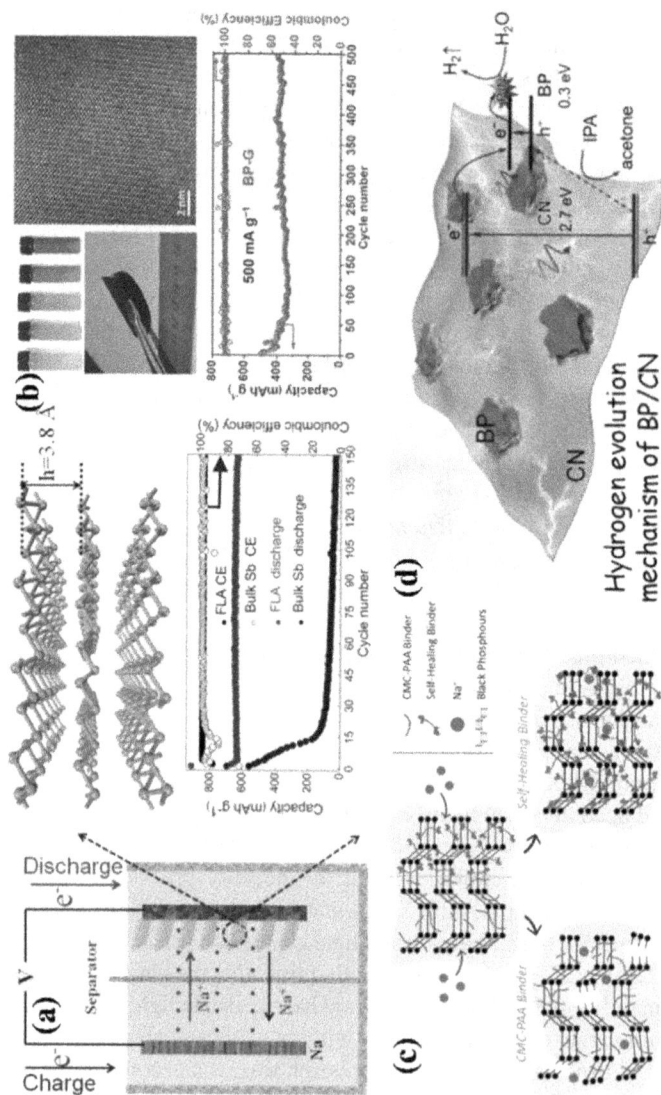

FIGURE 4.6 (a) Atomic structure diagram of few-layer antimonene, schematic diagram, and cycle stability of sodium ion half-cell based on few-layer antimonene. Reproduced with permission,[120] copyright 2018, American Chemical Society. (b) Photograph of liquid-phase exfoliated BP nanosheet solution and BP-G mixed paper; cycle stability of flexible lithium-ion battery based on BP-G hybrid paper electrode. Reproduced with permission,[115] copyright 2016, John Wiley & Sons. (c) Schematic diagram of self-healing BP sodium ion battery structure. Reproduced with permission,[117] copyright 2021, American Chemical Society. (d) Schematic diagram of the charge transfer process and mechanism of BP/g-CN photocatalytic hydrogen evolution. Reproduced with permission,[125] copyright 2019, John Wiley & Sons.

electron-hole pairs, thereby improving the photocatalytic performance of g-C_3N_4 (Figure 4.6(d)). Not only that, Jiang et al.[126] developed a simple strategy to prepare black phosphorous nanosheet films on titanium foil, and demonstrated good electrocatalytic oxygen evolution performance. Subsequently, our team reported that the electrocatalytic oxygen evolution performance of black phosphorus can be effectively enhanced by adjusting the number of layers of black phosphorus nanosheets[127] or using Au nanoparticles to modify the surface.[128] It is worth noting that the modification of Au nanoparticles can promote the transformation of black phosphorus from a semiconductor phase to a metal phase to effectively improve its conductivity, and it can effectively improve the adsorption of black phosphorus to electrolyte ions to show the Gibbs free energy that is close to the most valuable. In addition, Kokabi et al.[129] reported that the armchair edge of the single-layer antimonene nanosheets has high catalytic activity and photocatalytic performance in the entire pH range, which makes it a promising candidate material for photocatalytic water splitting. Som et al.[130] used the latest density functional theory (DFT) to study the catalytic activity of arsenic for hydrogen release reaction and oxygen release reaction, and confirmed that arsenic is a potential candidate for HER. And it was found that the HER activity of O-doped arsenene increased by 82%, and the OER activity of B-doped arsenene increased by 87%. In addition, we produced several layers of antimonene through the liquid stripping method and achieved optimal bifunctional electrocatalytic activity and structural stability.[131] Yang et al.[132] demonstrated the independent first simple large-scale synthesis method of bismuthene, and demonstrated its high electrocatalytic efficiency from the reduction of CO_2 to formic acid. Moreover, there are many reports that the photocatalytic performance of group VA 2D Xenes materials can be adjusted through various strategies, or as a co-catalyst to improve the performance of other catalysts. These reports confirm that group VA 2D Xenes materials have promising applications in the field of photocatalysis/electrocatalytic.

4.4.4 BIOMEDICINE

The exploration of group VA 2D Xenes materials in the biomedical field has become a hot topic due to their unique physical and chemical properties. They have good biocompatibility, especially P element, which itself is one of the trace elements required by the human body. First, group VA 2D Xenes materials are often used as drug carriers due to their unique two-dimensional structure and large specific surface area[25, 133, 134] (Figure 4.7(a)). Tao et al.[135] developed PEGylated BP nanosheets as an efficient loading platform for therapeutic drugs to treat the human body. They not only showed excellent biocompatibility, but also the doxorubicin-loaded PEGylated BP nanosheets showed enhanced anti-tumor effects in vitro and in vivo. In addition, the excellent light-to-heat conversion effect of group VA 2D Xenes materials is used to inactivate cancer cells. For example, Shao et al.[136] used poly(lactic-glycolic acid copolymer) loaded with black phosphorus quantum dots (BPQDs/PLGA nanospheres) for tumor treatment (Figure 4.7(c)). In vitro and in vivo experiments show that BPQDs/PLGA nanospheres have

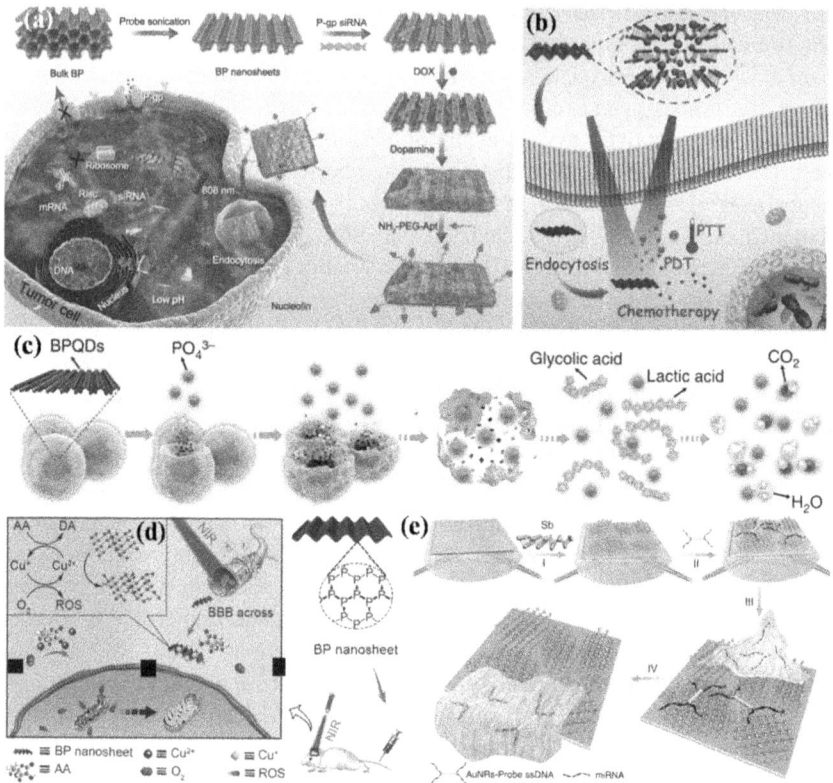

FIGURE 4.7 (a) Schematic illustration of the procedure used to fabricate nanostructures and the combined chemo/gene/photothermal targeted therapy of tumor cells. Reproduced with permission,[134] copyright 2019, John Wiley & Sons. (b) Chematic illustration of BP-based drug delivery system for synergistic photodynamic/photothermal/chemotherapy of cancer. Reproduced with permission,[137] copyright 2017, John Wiley & Sons. (c) Schematic representation of the degradation process of the BPQDs/PLGA NSs in the physiological environment. Reproduced with permission,[136] copyright 2016, *Springer Nature*. (d) Schematic illustration of BP nanosheets as blood–brain barrier. Reproduced with permission,[139] copyright 2018, John Wiley & Sons. (e) Fabrication of a miRNA sensor integrated with antimonene nanomaterials. Reproduced with permission,[147] copyright 2019, *Springer Nature*.

insignificant toxicity and good biocompatibility, and have excellent photothermal therapy efficiency and tumor targeting ability, and can efficiently ablate tumors under near-infrared laser irradiation. Subsequently, Chen et al.[137] proposed an interesting concept of collaborative photodynamic/photothermal/chemotherapy for cancer based on a drug delivery system based on BP nanosheets, as shown in Figure 4.7(b). The drug delivery system not only has improved drug loading efficiency, but also exhibits a significantly enhanced ability to kill tumor cells under the synergistic combination of chemotherapy, photothermal therapy, and photodynamic therapy. In addition, Guo[138] also reported that a multifunctional nano-platform based on black phosphorous quantum dots can be used not only for photothermal therapy and photodynamic therapy but also as a loading platform for fluorescent molecules to achieve reliable imaging of cancer cells. Chen et al.[139] also proved that 2D black phosphorus nanosheets can effectively and selectively capture Cu^{2+}, thereby protecting neuronal cells from Cu^{2+}-induced neurotoxicity, as shown in Figure 4.7(d). In addition to these, Wang et al.[140] found that the new two-dimensional arsenic nanosheets showed effective inhibition of NB4 promyelocytic leukemia cells and induced apoptosis, while being non-toxic to normal cells. At the same time, studies have shown that antimonene has excellent application prospects as a drug delivery platform[141] and NIR-induced tumor ablation.[142] Interestingly, Duo et al.[143] achieved cancer cell apoptosis and tumor activity regression by inducing the strong oxidative stress response of antimones and their significant high radiotoxicity in the body. Guo et al.[144] found that under manually controlled NIR irradiation, bismuth-based nanosheets simultaneously induced enhanced pathological permeability and retention of tumors, increased tumor-infiltrating lymphocytes (TIL) recruitment, and triggered programmed siRNA release, thereby amplifying anti-PD-L1 immunotherapy. Group VA 2D Xenes materials are often used as bioluminescent probes due to their excellent fluorescence effect and biocompatibility. Zhou et al.[145] used black phosphorous nanosheets as a fluorescence quenching material for the first time and developed a sensitive sensing platform for rapid detection of microRNA. Xue et al.[146] fundamentally explored the quenching mechanism of bismuth and sensitively detected miRNA molecules for early cancer. Xue et al.[147] developed a two-dimensional antimonene-based surface plasmon resonance sensor for specific label-free detection of clinically relevant biomarkers, providing a promising approach for early diagnosis, staging, and monitoring of cancer (Figure 4.7(e).

4.5 SUMMARY

The successful preparation and application of group VA 2D Xenes materials (2D phosphorene, arsenene, antimonene, and bismuthene) leads to the improvement and expansion of the field of 2D materials. In this chapter, we comprehensively reviewed the development history of group VA 2D Xenes materials and made a comprehensive summary and discussion on their properties, preparation methods, and application fields. Group VA 2D Xenes materials are semiconductors with a wide band gap ranging from 0.36 to 2.62 eV. Among them, arsenene, antimonene, and bismuthene can realize the transformation of metal phase and semiconductor phase through layer number control, which provides more abundant applications for 2D materials in the fields of optics and electronics. In addition, the outstanding anisotropy of Group VA 2D Xenes materials makes their optical, electronic, and thermodynamic properties more colorful, which greatly enriches their applications in various fields. At present, the research on the topological properties of group VA 2D Xenes materials is still at the stage of theoretical calculation and performance prediction, and the development of its practical applications will also be the direction of its further development. In addition, in view of the good application of group VA 2D Xenes materials in photodetectors, photo/electrocatalysis, and other fields, it is predicted that they have potential applications in the fields of solar cells, and gas and chemical sensors. The content in this chapter will help readers to fully understand gtroup VA 2D Xenes materials and provide theoretical guidance for their application in various new high-performance electronic and optoelectronic devices.

REFERENCES

1. Zhang, S. et al., Recent progress in 2D group-VA semiconductors: from theory to experiment. *Chemical Society Reviews*, 47, 3, 2018.
2. Li, L. et al., Black phosphorus field-effect transistors. *Nature Nanotechnology*, 9, 5, 2014.
3. Sun, Z. et al., Ultrasmall black phosphorus quantum dots: synthesis and use as photothermal agents. *Angewandte Chemie International Edition*, 54, 39, 2015.
4. Rodin, A., Carvalho, A. & Neto, A. C., Strain-induced gap modification in black phosphorus. *Physical Review Letters*, 112, 17, 2014.
5. Deng, B. et al., Efficient electrical control of thin-film black phosphorus bandgap. *Nature Communications*, 8, 1, 2017.
6. Tran, V., Soklaski, R., Liang, Y. & Yang, L., Layer-controlled band gap and anisotropic excitons in few-layer black phosphorus. *Physical Review B*, 89, 23, 2014.
7. Su, S. H. et al., Band gap - tunable molybdenum sulfide selenide monolayer alloy. *Small*, 10, 13, 2014.
8. Qiao, J., Kong, X., Hu, Z.-X., Yang, F. & Ji, W., High-mobility transport anisotropy and linear dichroism in few-layer black phosphorus. *Nature Communications*, 5, 1, 2014.
9. Qiao, H. et al., Self-powered photodetectors based on 0D/2D mixed dimensional heterojunction with black phosphorus quantum dots as hole accepters. *Applied Materials Today*, 20, 2020.
10. Qiao, H. et al., Tunable electronic and optical properties of 2D monoelemental materials beyond graphene for promising applications. *Energy & Environmental Materials*, 4, 4, 2020.
11. Hu, R. et al., Recent advances of monoelemental 2D materials for photoctalytic applications. *Journal of Hazardous Materials*, 405, 124179, 2020.
12. Pumera, M. & Sofer, Z., 2D monoelemental arsenene, antimonene, and bismuthene: beyond black phosphorus. *Advanced Materials*, 29, 21, 2017.
13. Huang, Z. et al., Structures, properties and application of 2D monoelemental materials (Xenes) as graphene analogues under defect engineering. *Nano Today*, 35, 2020.
14. Zhu, S.-Y. et al., Evidence of topological edge states in buckled antimonene monolayers. *Nano Letters*, 19, 9, 2019.
15. Radha, S. K. & Lambrecht, W. R., Topological band structure transitions and goniopolar transport in honeycomb antimonene as a function of buckling. *Physical Review B*, 101, 23, 2020.
16. Kadioglu, Y. et al., Modification of electronic structure, magnetic structure, and topological phase of bismuthene by point defects. *Physical Review B*, 96, 24, 2017.
17. Xia, F., Wang, H. & Jia, Y., Rediscovering black phosphorus as an anisotropic layered material for optoelectronics and electronics. *Nature Communications*, 5, 1, 2014.
18. Wang, X. et al., Highly anisotropic and robust excitons in monolayer black phosphorus. *Nature Nanotechnology*, 10, 6, 2015.

19. Tao, J. et al., Mechanical and electrical anisotropy of few-layer black phosphorus. *ACS Nano*, 9, 11, 2015.
20. Yin, Y. et al., Anisotropic transport property of antimonene MOSFETs. *ACS Applied Materials & Interfaces*, 12, 19, 2020.
21. Zhou, J. et al., Few-layer bismuthene with anisotropic expansion for high-areal-capacity sodium-ion batteries. *Advanced Materials*, 31, 12, 2019.
22. Zeraati, M., Allaei, S. M. V., Sarsari, I. A., Pourfath, M. & Donadio, D., Highly anisotropic thermal conductivity of arsenene: an ab initio study. *Physical Review B*, 93, 8, 2016.
23. Liu, G. et al., Black phosphorus nanosheets-based stable drug delivery system via drug-self-stabilization for combined photothermal and chemo cancer therapy. *Chemical Engineering Journal*, 375, 2019.
24. Tao, W. et al., Cancer theranostics: two-dimensional antimonene-based photonic nanomedicine for cancer theranostics (Adv. Mater. 38/2018). *Advanced Materials*, 30, 38, 2018.
25. Tao, W. et al., Emerging two-dimensional monoelemental materials (Xenes) for biomedical applications. *Chemical Society Reviews*, 48, 11, 2019.
26. Bridgman, P., Two new modifications of phosphorus. *Journal of the American Chemical Society*, 36, 7, 1914.
27. Peng, B. et al., The conflicting role of buckled structure in phonon transport of 2D group-IV and group-V materials. *Nanoscale*, 9, 22, 2017.
28. Zhang, S. et al., Semiconducting group 15 monolayers: a broad range of band gaps and high carrier mobilities. *Angewandte Chemie*, 128, 5, 2016.
29. Aktürk, E., Aktürk, O. Ü. & Ciraci, S., Single and bilayer bismuthene: stability at high temperature and mechanical and electronic properties. *Physical Review B*, 94, 1, 2016.
30. Zhang, S., Yan, Z., Li, Y., Chen, Z. & Zeng, H., Atomically thin arsenene and antimonene: semimetal–semiconductor and indirect–direct band-gap transitions. *Angewandte Chemie*, 127, 10, 2015.
31. Li, L. et al., Black phosphorus field-effect transistors. *Nature Nanotechnology*, 9, 5, 2014.
32. Pizzi, G. et al., Performance of arsenene and antimonene double-gate MOSFETs from first principles. *Nature Communications*, 7, 1, 2016.
33. Çakır, D., Sahin, H. & Peeters, F. M., Tuning of the electronic and optical properties of single-layer black phosphorus by strain. *Physical Review B*, 90, 20, 2014.
34. Low, T. et al., Tunable optical properties of multilayer black phosphorus thin films. *Physical Review B*, 90, 7, 2014.
35. Shu, H., Li, Y., Niu, X. & Guo, J., Electronic structures and optical properties of arsenene and antimonene under strain and an electric field. *Journal of Materials Chemistry C*, 6, 1, 2018.
36. Xie, M. et al., van der Waals bilayer antimonene: a promising thermophotovoltaic cell material with 31% energy conversion efficiency. *Nano Energy*, 38, 2017.
37. Mao, N. et al., Optical anisotropy of black phosphorus in the visible regime. *Journal of the American Chemical Society*, 138, 1, 2016.
38. Zhang, B., Zhang, H., Lin, J. & Cheng, X., First-principle study of seven allotropes of arsenene and antimonene: thermodynamic, electronic and optical properties. *Physical Chemistry Chemical Physics*, 20, 48, 2018.
39. Aierken, Y., Çakır, D., Sevik, C. & Peeters, F. M., Thermal properties of black and blue phosphorenes from a first-principles quasiharmonic approach. *Physical Review B*, 92, 8, 2015.
40. Zhu, L., Zhang, G. & Li, B., Coexistence of size-dependent and size-independent thermal conductivities in phosphorene. *Physical Review B*, 90, 21, 2014.

41. Sun, B. et al., Temperature dependence of anisotropic thermal-conductivity tensor of bulk black phosphorus. *Advanced Materials*, 29, 3, 2017.
42. Jang, H., Wood, J. D., Ryder, C. R., Hersam, M. C. & Cahill, D. G., Anisotropic thermal conductivity of exfoliated black phosphorus. *Advanced Materials*, 27, 48, 2015.
43. Luo, Z. et al., Anisotropic in-plane thermal conductivity observed in few-layer black phosphorus. *Nature Communications*, 6, 1, 2015.
44. Lee, S. et al., Anisotropic in-plane thermal conductivity of black phosphorus nanoribbons at temperatures higher than 100 K. *Nature Communications*, 6, 1, 2015.
45. Wang, S., Wang, W. & Zhao, G., Thermal transport properties of antimonene: an ab initio study. *Physical Chemistry Chemical Physics*, 18, 45, 2016.
46. Liu, H. et al., Phosphorene: an unexplored 2D semiconductor with a high hole mobility. *ACS Nano*, 8, 4, 2014.
47. Yasaei, P. et al., High-quality black phosphorus atomic layers by liquid-phase exfoliation. *Advanced Materials*, 27, 11, 2015.
48. Brent, J. R. et al., Production of few-layer phosphorene by liquid exfoliation of black phosphorus. *Chemical Communications*, 50, 87, 2014.
49. Hanlon, D. et al., Liquid exfoliation of solvent-stabilized few-layer black phosphorus for applications beyond electronics. *Nature Communications*, 6, 1, 2015.
50. Beladi-Mousavi, S. M., Pourrahimi, A. M., Sofer, Z. & Pumera, M., Atomically thin 2D-arsenene by liquid-phased exfoliation: toward selective vapor sensing. *Advanced Functional Materials*, 29, 5, 2019.
51. Gibaja, C. et al., Few-layer antimonene by liquid-phase exfoliation. *Angewandte Chemie*, 128, 46, 2016.
52. Zhang, W. et al., Liquid-phase exfoliated ultrathin Bi nanosheets: uncovering the origins of enhanced electrocatalytic CO2 reduction on two-dimensional metal nanostructure. *Nano Energy*, 53, 2018.
53. Ambrosi, A., Sofer, Z. & Pumera, M., Electrochemical exfoliation of layered black phosphorus into phosphorene. *Angewandte Chemie*, 129, 35, 2017.
54. Erande, M. B., Pawar, M. S. & Late, D. J., Humidity sensing and photodetection behavior of electrochemically exfoliated atomically thin-layered black phosphorus nanosheets. *ACS Applied Materials & Interfaces*, 8, 18, 2016.
55. Marzo, A. M. L., Gusmão, R., Sofer, Z. & Pumera, M., Towards antimonene and 2D antimony telluride through electrochemical exfoliation. *Chemistry–A European Journal*, 2020.
56. Kovalska, E., Antonatos, N., Luxa, J. & Sofer, Z. k., "Top-down" arsenene production by low-potential electrochemical exfoliation. *Inorganic Chemistry*, 59, 16, 2020.
57. Smith, J. B., Hagaman, D. & Ji, H.-F., Growth of 2D black phosphorus film from chemical vapor deposition. *Nanotechnology*, 27, 21, 2016.
58. Wu, Q. & Song, Y. J., The environmental stability of large-size and single-crystalline antimony flakes grown by chemical vapor deposition on SiO 2 substrates. *Chemical Communications*, 54, 69, 2018.
59. Ji, J. et al., Two-dimensional antimonene single crystals grown by van der Waals epitaxy. *Nature Communications*, 7, 1, 2016.
60. Lu, D. et al., Strong temperature-strain coupling in the interface of Sb thin film on flexible PDMS substrate. *Applied Physics Letters*, 115, 12, 2019.
61. Zhou, Q. et al., Self-powered ultra-broadband and flexible photodetectors based on the bismuth films by vapor deposition. *ACS Applied Electronic Materials*, 2, 5, 2020.

62. Kuriakose, S. et al., Monocrystalline antimonene nanosheets via physical vapor deposition. *Advanced Materials Interfaces*, 7, 24, 2020.
63. Yang, Z. et al., Field-effect transistors based on amorphous black phosphorus ultra-thin films by pulsed laser deposition. *Advanced Materials*, 27, 25, 2015.
64. Wu, X. et al., Epitaxial growth and air-stability of monolayer antimonene on PdTe2. *Advanced Materials*, 29, 11, 2017.
65. Chen, H.-A. et al., Single-crystal antimonene films prepared by molecular beam epitaxy: selective growth and contact resistance reduction of the 2D material heterostructure. *ACS Applied Materials & Interfaces*, 10, 17, 2018.
66. Zhong, W. et al., Anisotropic thermoelectric effect and field-effect devices in epitaxial bismuthene on Si (111). *Nanotechnology*, 31, 47, 2020.
67. Du, Y., Liu, H., Deng, Y. & Ye, P. D., Device perspective for black phosphorus field-effect transistors: contact resistance, ambipolar behavior, and scaling. *ACS Nano*, 8, 10, 2014.
68. Perello, D. J., Chae, S. H., Song, S. & Lee, Y. H., High-performance n-type black phosphorus transistors with type control via thickness and contact-metal engineering. *Nature Communications*, 6, 1, 2015.
69. Na, J. et al., Few-layer black phosphorus field-effect transistors with reduced current fluctuation. *ACS Nano*, 8, 11, 2014.
70. Wood, J. D. et al., Effective passivation of exfoliated black phosphorus transistors against ambient degradation. *Nano Letters*, 14, 12, 2014.
71. He, D. et al., High-performance black phosphorus field-effect transistors with long-term air stability. *Nano Letters*, 19, 1, 2018.
72. Yin, D., AlMutairi, A. & Yoon, Y., Assessment of high-frequency performance limit of black phosphorus field-effect transistors. *IEEE Transactions on Electron Devices*, 64, 7, 2017.
73. Wang, H. et al., Black phosphorus radio-frequency transistors. *Nano Letters*, 14, 11, 2014.
74. Zhu, W. et al., Flexible black phosphorus ambipolar transistors, circuits and AM demodulator. *Nano Letters*, 15, 3, 2015.
75. Li, H., Xu, P., Xu, L., Zhang, Z. & Lu, J., Negative capacitance tunneling field effect transistors based on monolayer arsenene, antimonene, and bismuthene. *Semiconductor Science and Technology*, 34, 8, 2019.
76. Tian, H. et al., A dynamically reconfigurable ambipolar black phosphorus memory device. *ACS Nano*, 10, 11, 2016.
77. Lee, D. et al., Black phosphorus nonvolatile transistor memory. *Nanoscale*, 8, 17, 2016.
78. Liu, Q., Zhang, X., Abdalla, L., Fazzio, A. & Zunger, A., Switching a normal insulator into a topological insulator via electric field with application to phosphorene. *Nano Letters*, 15, 2, 2015.
79. Zhang, H., Ma, Y. & Chen, Z., Quantum spin hall insulators in strain-modified arsenene. *Nanoscale*, 7, 45, 2015.
80. Zhao, M., Zhang, X. & Li, L., Strain-driven band inversion and topological aspects in Antimonene. *Scientific Reports*, 5, 1, 2015.
81. Song, Z. et al., Quantum spin Hall insulators and quantum valley Hall insulators of BiX/SbX (X= H, F, Cl and Br) monolayers with a record bulk band gap. *NPG Asia Materials*, 6, 12, 2014.
82. Wu, J. et al., Colossal ultraviolet photoresponsivity of few-layer black phosphorus. *ACS Nano*, 9, 8, 2015.
83. Buscema, M. et al., Fast and broadband photoresponse of few-layer black phosphorus field-effect transistors. *Nano Letters*, 14, 6, 2014.

84. Low, T., Engel, M., Steiner, M. & Avouris, P., Origin of photoresponse in black phosphorus phototransistors. *Physical Review B*, 90, 8, 2014.
85. Huang, M. et al., Broadband black-phosphorus photodetectors with high responsivity. *Advanced Materials*, 28, 18, 2016.
86. Guo, Q. et al., Black phosphorus mid-infrared photodetectors with high gain. *Nano Letters*, 16, 7, 2016.
87. Huang, L. et al., Waveguide-integrated black phosphorus photodetector for mid-infrared applications. *ACS Nano*, 13, 1, 2018.
88. Youngblood, N., Chen, C., Koester, S. J. & Li, M., Waveguide-integrated black phosphorus photodetector with high responsivity and low dark current. *Nature Photonics*, 9, 4, 2015.
89. Chen, X. et al., Widely tunable black phosphorus mid-infrared photodetector. *Nature Communications*, 8, 1, 2017.
90. Na, J., Park, K., Kim, J. T., Choi, W. K. & Song, Y.-W., Air-stable few-layer black phosphorus phototransistor for near-infrared detection. *Nanotechnology*, 28, 8, 2017.
91. Zhang, Y. et al., Self-healable black phosphorus photodetectors. *Advanced Functional Materials*, 29, 49, 2019.
92. Xiao, Q. et al., Antimonene-based flexible photodetector. *Nanoscale Horizons*, 5, 1, 2020.
93. Ren, X. et al., Environmentally robust black phosphorus nanosheets in solution: application for self-powered photodetector. *Advanced Functional Materials*, 27, 18, 2017.
94. Huang, H. et al., Two-dimensional bismuth nanosheets as prospective photo-detector with tunable optoelectronic performance. *Nanotechnology*, 29, 23, 2018.
95. Wang, B. et al., Photoelectrochemical self-powered photodetector based on 2D liquid-exfoliated bismuth nanosheets: with novel structures for portability and flexibility. *Materials Today Nano*, 14, 2021.
96. Su, L. et al., Halogenated antimonene: one-step synthesis, structural simulation, tunable electronic and photoresponse property. *Advanced Functional Materials*, 29, 45, 2019.
97. Xing, C. et al., Ultrasmall bismuth quantum dots: facile liquid-phase exfoliation, characterization, and application in high-performance UV–Vis photodetector. *Acs Photonics*, 5, 2, 2018.
98. Li, D. et al., Polarization and thickness dependent absorption properties of black phosphorus: new saturable absorber for ultrafast pulse generation. *Scientific Reports*, 5, 1, 2015.
99. Jiang, X.-F. et al., Tunable broadband nonlinear optical properties of black phosphorus quantum dots for femtosecond laser pulses. *Materials*, 10, 2, 2017.
100. Du, L. et al., Broadband nonlinear optical response of single-crystalline bismuth thin film. *ACS Applied Materials & Interfaces*, 11, 39, 2019.
101. Lu, L. et al., Broadband nonlinear optical response in few-layer antimonene and antimonene quantum dots: a promising optical kerr media with enhanced stability. *Advanced Optical Materials*, 5, 17, 2017.
102. Zhang, S. et al., Extraordinary photoluminescence and strong temperature/angle-dependent Raman responses in few-layer phosphorene. *ACS Nano*, 8, 9, 2014.
103. Chen, C. et al., Bright mid-infrared photoluminescence from thin-film black phosphorus. *Nano Letters*, 19, 3, 2019.
104. Chang, T.-Y. et al., Black phosphorus mid-infrared light-emitting diodes integrated with silicon photonic waveguides. *Nano Letters*, 20, 9, 2020.
105. Tsai, H.-S. et al., Direct synthesis and practical bandgap estimation of multilayer arsenene nanoribbons. *Chemistry of Materials*, 28, 2, 2016.

106. Tsai, H.-S., Chen, C.-W., Hsiao, C.-H., Ouyang, H. & Liang, J.-H., The advent of multilayer antimonene nanoribbons with room temperature orange light emission. *Chemical Communications*, 52, 54, 2016.

107. Liu, Y. et al., Antimonene quantum dots as an emerging fluorescent nanoprobe for the pH-mediated dual-channel detection of tetracyclines. *Small*, 16, 42, 2020.

108. Hussain, N. et al., Ultrathin Bi nanosheets with superior photoluminescence. *Small*, 13, 36, 2017.

109. Li, W., Yang, Y., Zhang, G. & Zhang, Y.-W., Ultrafast and directional diffusion of lithium in phosphorene for high-performance lithium-ion battery. *Nano Letters*, 15, 3, 2015.

110. Wu, Z.-S., Ren, W., Xu, L., Li, F. & Cheng, H.-M., Doped graphene sheets as anode materials with superhigh rate and large capacity for lithium ion batteries. *ACS Nano*, 5, 7, 2011.

111. Persson, K., Hinuma, Y., Meng, Y. S., van der Ven, A. & Ceder, G., Thermodynamic and kinetic properties of the Li-graphite system from first-principles calculations. *Physical Review B*, 82, 12, 2010.

112. Kulish, V. V., Malyi, O. I., Persson, C. & Wu, P., Phosphorene as an anode material for Na-ion batteries: a first-principles study. *Physical Chemistry Chemical Physics*, 17, 21, 2015.

113. Liu, X., Wen, Y., Chen, Z., Shan, B. & Chen, R., A first-principles study of sodium adsorption and diffusion on phosphorene. *Physical Chemistry Chemical Physics*, 17, 25, 2015.

114. Chen, L. et al., Scalable clean exfoliation of high-quality few-layer black phosphorus for a flexible lithium ion battery. *Advanced Materials*, 28, 3, 2016.

115. Sultana, I., Rahman, M. M., Ramireddy, T., Chen, Y. & Glushenkov, A. M., High capacity potassium-ion battery anodes based on black phosphorus. *Journal of Materials Chemistry A*, 5, 45, 2017.

116. Callegari, D. et al., Autonomous self-healing strategy for stable sodium-ion battery: a case study of black phosphorus anodes. *ACS Applied Materials & Interfaces*, 13, 11, 2021.

117. Benzidi, H. et al., Arsenene monolayer as an outstanding anode material for (Li/Na/Mg)-ion batteries: density functional theory. *Physical Chemistry Chemical Physics*, 21, 36, 2019.

118. Ye, X.-J., Zhu, G.-L., Liu, J., Liu, C.-S. & Yan, X.-H., Monolayer, bilayer, and heterostructure arsenene as potential anode materials for magnesium-ion batteries: a first-principles study. *The Journal of Physical Chemistry C*, 123, 25, 2019.

119. Tian, W. et al., Few-layer antimonene: anisotropic expansion and reversible crystalline-phase evolution enable large-capacity and long-life Na-ion batteries. *ACS Nano*, 12, 2, 2018.

120. Mao, X., Zhu, L. & Fu, A., Arsenene, antimonene and bismuthene as anchoring materials for lithium-sulfur batteries: a computational study. *International Journal of Quantum Chemistry*, 121, 14, 2020.

121. Singh, D. et al., Antimonene allotropes α-and β-phases as promising anchoring materials for lithium–sulfur batteries. *Energy & Fuels*, 35, 10, 2021.

122. Ren, X. et al., Single cobalt atom anchored black phosphorous nanosheets as an effective cocatalyst promotes photocatalysis. *ChemCatChem*, 12, 15, 2020.

123. Zhu, X. et al., Black phosphorus revisited: a missing metal-free elemental photocatalyst for visible light hydrogen evolution. *Advanced Materials*, 29, 17, 2017.

124. Wen, M. et al., A low-cost metal-free photocatalyst based on black phosphorus. *Advanced Science*, 6, 1, 2019.

125. Jiang, Q. et al., Facile synthesis of black phosphorus: an efficient electrocatalyst for the oxygen evolving reaction. *Angewandte Chemie*, 128, 44, 2016.
126. Ren, X. et al., Few-layer black phosphorus nanosheets as electrocatalysts for highly efficient oxygen evolution reaction. *Advanced Energy Materials*, 7, 19, 2017.
127. Qiao, H. et al., Black phosphorus nanosheets modified with Au nanoparticles as high conductivity and high activity electrocatalyst for oxygen evolution reaction. *Advanced Energy Materials*, 10, 44, 2020.
128. Kokabi, A. & Touski, S. B., Electronic and photocatalytic properties of Antimonene nanosheets. *Physica E: Low-dimensional Systems and Nanostructures*, 124, 2020.
129. Som, N. N., Mankad, V. & Jha, P. K., Hydrogen evolution reaction: the role of arsenene nanosheet and dopant. *International Journal of Hydrogen Energy*, 43, 47, 2018.
130. Ren, X. et al., Few-layer antimonene nanosheet: a metal-free bifunctional electrocatalyst for effective water splitting. *ACS Applied Energy Materials*, 2, 7, 2019.
131. Yang, F. et al., Bismuthene for highly efficient carbon dioxide electroreduction reaction. *Nature Communications*, 11, 1, 2020.
132. Liu, W., Dong, A., Wang, B. & Zhang, H., Current advances in black phosphorus-based drug delivery systems for cancer therapy. *Advanced Science*, 8, 5, 2021.
133. Zeng, X. et al., Polydopamine-modified black phosphorous nanocapsule with enhanced stability and photothermal performance for tumor multimodal treatments. *Advanced Science*, 5, 10, 2018.
134. Tao, W. et al., Black phosphorus nanosheets as a robust delivery platform for cancer theranostics. *Advanced Materials*, 29, 1, 2017.
135. Shao, J. et al., Biodegradable black phosphorus-based nanospheres for in vivo photothermal cancer therapy. *Nature Communications*, 7, 1, 2016.
136. Chen, W. et al., Black phosphorus nanosheet-based drug delivery system for synergistic photodynamic/photothermal/chemotherapy of cancer. *Advanced Materials*, 29, 5, 2017.
137. Li, Y. et al., Multifunctional nanoplatform based on black phosphorus quantum dots for bioimaging and photodynamic/photothermal synergistic cancer therapy. *ACS Applied Materials & Interfaces*, 9, 30, 2017.
138. Chen, W. et al., Black phosphorus nanosheets as a neuroprotective nanomedicine for neurodegenerative disorder therapy. *Advanced Materials*, 30, 3, 2018.
139. Wang, X. et al., Arsenene: a potential therapeutic agent for acute promyelocytic leukaemia cells by acting on nuclear proteins. *Angewandte Chemie International Edition*, 59, 13, 2020.
140. Tao, W. et al., Two-dimensional antimonene-based photonic nanomedicine for cancer theranostics. *Advanced Materials*, 30, 38, 2018.
141. Tao, W. et al., Antimonene quantum dots: synthesis and application as near-infrared photothermal agents for effective cancer therapy. *Angewandte Chemie*, 129, 39, 2017.
142. Duo, Y. et al., Ultraeffective cancer therapy with an antimonene-based X-ray radiosensitizer. *Advanced Functional Materials*, 30, 4, 2020.
143. Guo, M. et al., Few-layer bismuthene for checkpoint knockdown enhanced cancer immunotherapy with rapid clearance and sequentially triggered one-for-all strategy. *ACS nano*, 14, 11, 2020.
144. Zhou, J. et al., Black phosphorus nanosheets for rapid microRNA detection. *Nanoscale*, 10, 11, 2018.
145. Xue, T. et al., Ultrasensitive detection of microRNA using a bismuthene-enabled fluorescence quenching biosensor. *Chemical Communications*, 56, 51, 2020.
146. Xue, T. et al., Ultrasensitive detection of miRNA with an antimonene-based surface plasmon resonance sensor. *Nature Communications*, 10, 1, 2019.

5 Group VIA of 2D Xenes materials

Xiong-Xiong Xue and Yexin Feng

CONTENTS

5.1 INTRODUCTION

Besides group-IIIA, -IVA, and -VA elemental 2D Xenes, people have also been working on developing group VIA elemental 2D materials to further enrich the family of 2D materials. As one of group VIA chalcogens, elemental tellurium (Te) ordinarily forms a chiral crystal consisting of helical chains parallel to one of the crystal lattices, as shown in Figure 5.1(a), which is the most stable configuration at ambient conditions. In other words, bulk Te is essentially a bundle of weakly interacting one-dimensional nanowires, in which each atom only bonds to the two nearest atoms with two-fold coordination. This particular helical structure endows elemental Te with a strong tendency to form one-dimensional (1D)

DOI: 10.1201/9781003207122-5

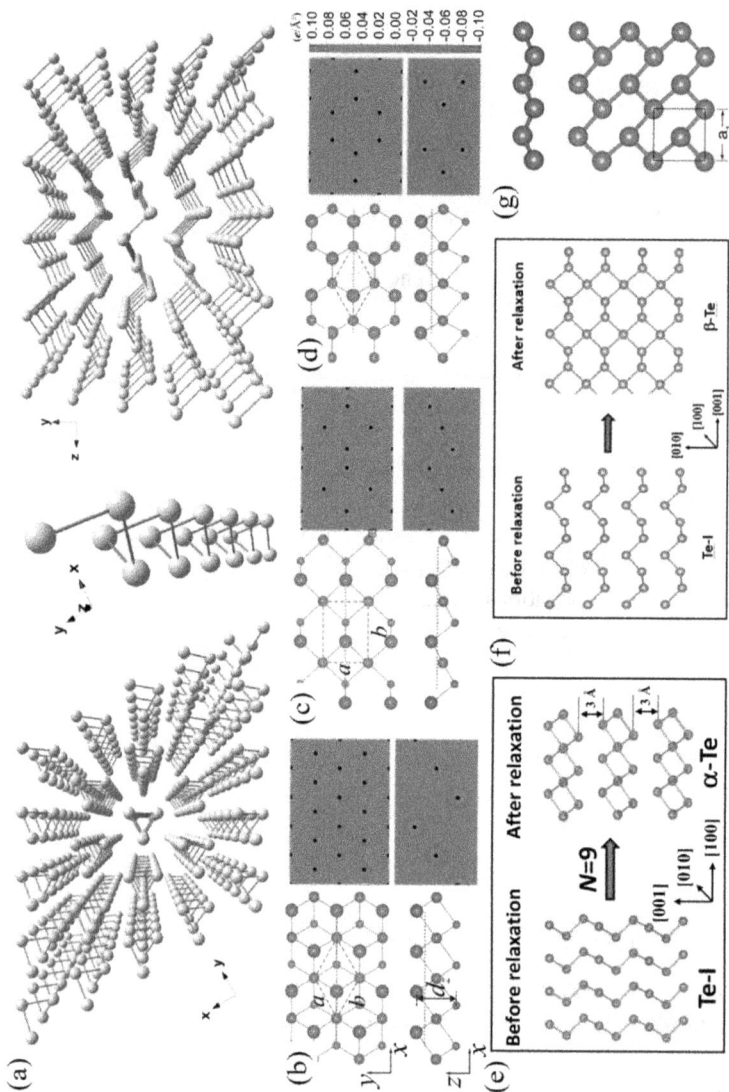

FIGURE 5.1 (a) Crystal structure of bulk tellurium viewed from the z-axis, single-molecular chain, and the x-axis.[27] (b)–(c) Top and side views of the atomic structures and total charge densities of α-Te, β-Te, and γ-Te. (e), (f) The geometric structures of Te slabs with magic thickness of N = 9 and truncated along the [100] direction of bulk Te-I before and after structural optimization.[1] (g) Atomic structure of square tellurene.[3]

nanoarchitectures, including nanowires, nanorods, and nanotubes, which means a single layer of Te atoms (Tellurene) is difficult to obtain by means of simple mechanical exfoliating like graphene from graphite. Therefore, most previous studies about Te-based materials focused on 1D nanoarchitectures. It was until 2017, when Zhu et al. successfully predicted and prepared 2D Te monolayer named tellurene,[1] which further stimulated people's interest in studying 2D group VIA Xenes. Inspired by the discovery of tellurene, people have also successfully demonstrated a serious of 2D elemental Se monolayers named selenene.[2] Since then, research on these two newest members of 2D elemental materials has recently become a hot topic. A lot of theoretical and experimental studies have demonstrated that they possess intriguing structural and electronic properties, such as excellent environment stability, high photoconductivity, high carrier mobility, and thermoelectric response, together with some appealing applications including high-performance photodetectors, field-effect transistors, and thermoelectric devices.

In this chapter, we aim to summarize these structural and electronic properties, synthesis methods, and potential nanoelectronics applications of group VIA 2D elemental materials: tellurene and selenene. In each section, we will begin with crystal structures of allotropes and the experimental synthesis process before the review of physical properties including basic electronic structures and the resulting electrical properties such as thermoelectricity and ferroelectricity, etc. Then, we will highlight some promising device applications: field-effect transistors, photodetectors, and optical modulators. Finally, we will present the outlook of prospective challenges and further development of group VIA 2D elemental materials.

5.2 TELLURENE

5.2.1 CRYSTAL STRUCTURE OF TELLURENE ALLOTROPES

By combining the state-of-the-art global structural searching with first-principles density functional theory (DFT) calculations, Zhu et al. successfully predicted a new category of 2D group VIA Te monolayers named tellurene, which could be stabilized in three different phases α-, β-, and γ-Te.[1] As shown in Figures 1(b) and (d), α- and γ-Te show respectively 1T and 2H structures commonly adopted by 2D transition metal dichalcogenides MX_2 (M and X denote transition metal and chalcogen, respectively). Tetragonal β-Te consists of the planar zigzag-like four-membered and armchair-like six-membered rings arranged alternately from the top view. The behind formation mechanism of tellurene is rooted in the inherent multivalent nature of Te with metal- and semiconductor-like valance states. α- and γ-Te phases exhibit a mixture of threefold and sixfold coordination characteristics, while β-Te phase shows a mixture of threefold and fourfold coordination characteristics. Total charge densities and Bader charge analysis demonstrated that for α-, β-, and γ-Te, the central atoms interact with outer atoms in metal-ligand-like bonding, in which the central-layer and outer-layer Te atoms contain metal- and semiconductor-like bonding characteristics, respectively. In particular, for γ-Te, there also exists the σ bond between the outer atoms. It was also revealed that α-Te monolayers or multilayers could be spontaneously generated from magic thicknesses that are divisible evenly by three layers along the [001] direction of bulk Te (see Figure 5.1(e)). In contrast to the thickness-dependent structural phase transition mechanism of α-Te, the formation of β-Te phase is the natural consequence of structural relaxation. After relaxation, the helical chains in thin films at proper thicknesses become closer to each other, finally forming the β-Te phase, as shown in Figure 5.1(f). In addition, Xian et al. also predicted another novel stable 2D Te layered structure, namely square tellurene in the same year, as shown in Figure 5.1(g).[3] Similar to other 2D elemental materials such as silicene and germanene with a square unit, the square tellurene exhibit chair-like bucked configuration. This square phase was proposed to be potentially grown on proper substrates, such as Au(100) surface.

5.2.2   Synthesis Methods

5.2.2.1   Molecular Beam Epitaxy

The molecular beam epitaxy (MBE) method can produce high-quality 2D layered materials and provide atomic-level precision control on the thickness, which is beneficial for the fundamental learning of 2D elementary materials. In addition to the theoretical prediction of 2D Te structures, Zhu et al. have also successfully grown β-Te film on highly oriented pyrolytic graphite (HOPG) substrate by using MBE.[1] Soon afterwards, Chen et al. also applied MBE technology to grow theoretically predicted β-Te films on HOPG substrate, and scanning tunneling spectroscopy measurements revealed the semiconducting properties of tellurene films.[4] As shown in Figure 5.2(a), the atomically resolved scanning tunneling microscopy (STM) topographic image clearly indicated that flat Te thin film showed a rectangular unit cell. The measured unit cell area and step height matched very well with the calculated structural parameters of the β-Te phase. Furthermore, Guo et al. obtained monolayer and few-layer tellurium films by MBE on a graphene/6H-SiC(0001) substrate and found that the band gap of Te films decreased monotonically with the increase of thickness, up to the near-infrared band for monolayer film.[5] Although MBE combined with STM could grow and in situ characterize Te films, it suffers from poor efficiency, high cost, and limitations of substrate.

5.2.2.2   Physical and Chemical Vapor Deposition

Vapor deposition should be one of the potential methods to prepare large-area and high-quality 2D thin film materials. In addition, it has been proven that Physical Vapor Deposition (PVD) method can grow 2D tellurene. The PVD technique commonly requires to heat the target source reservoir to control the deposition of materials onto the substrates. Recently, using the horizontal PVD method, Yang et al. synthesized tellurene sheets in the trigonal layered form on a variety of

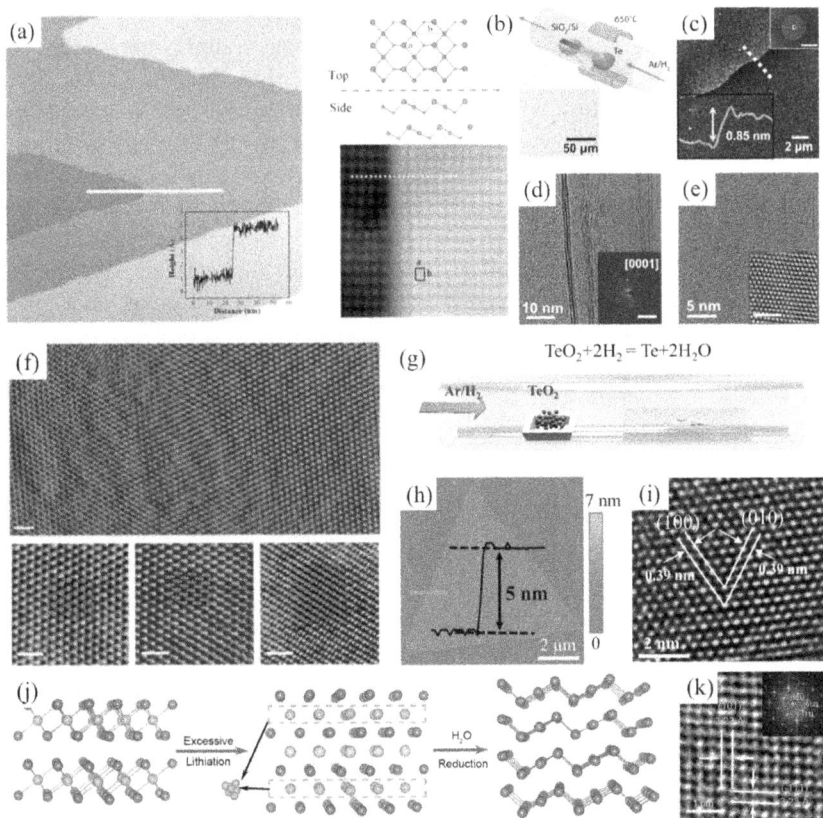

FIGURE 5.2 (a) Topographic image of an epitaxial Te film, the top and side views of a stick-and-ball model of the β-Te layer, and an atomic resolution STM image.[4] (b) Schematic diagram of vapor deposition of tellurene on SiO_2/Si substrate. (c) Atomic force microscopy (AFM) image and height profile. (d) Transmission electron microscopy (TEM) image of Te flake showing hexagonal spots in the selected-area electron diffraction pattern. (e) High-resolution HRTEM of the indicated region. (f) High resolution high angle annular darkfield scanning transmission electron microscopy (HAADF-STEM) images of the rippled tellurene flake and atomically resolved HAADF-STEM images of the three tellurene polymorphs.[7] (g) Schematic diagram of the chemical vapor deposition method. (h) AFM image and height profile of a Te flake. (i) HRTEM image of the flake.[8] (j) Scheme of the topotactically chemical transformation for 2D Te nanosheets synthesis. (k) HRTEM images of Te nanosheets.[11]

surfaces of 2D layered crystals and found that the highly crystalline and self-oriented tellurene sheets with controlled anisotropy can only be obtained on GaS and GaSe surfaces.[6] Cross-sectional TEM and first-principles calculations suggested that the well-oriented tellurene chains along the GaSe armchair lattice direction arose from the strong degree of coupling between adjacent layers and the reduced total energy of the system. The angle-resolved Raman measurements revealed a remarkable structural anisotropy for synthesized tellurene sheets. Apte et al. also demonstrated synthesis of atomically thin tellurium films through the PVD method and investigated the polytypism in ultrathin films.[7] In addition, the larger-area films could be fabricated by another complimentary route of pulsed laser deposition (PLD). As shown in Figure 5.2(b), during the PVD-synthesis process, the thermal evaporation of bulk Te in an Ar/H_2 environment at the temperature of 650 °C was deposited on Si/SiO_2 substrates and the ultrathin tellurium films were obtained after cooling down. The atomic force microscopy image in Figure 5.2(c) revealed that the synthesized tellurium films had the thickness of 0.85 nm, corresponding to three atomic layers. In Figure 5.2(d), the TEM image of selected-area electron diffraction pattern confirmed the hexagonal symmetry of tellurium films with three different sets of six-fold diffraction spots, which indicated the turbostratic disorder between three layers. The high-resolution STEM image displayed in Figure 5.2(f) further revealed the coexistance of all three theoretically predicted polytypes of α-, β-, and γ-Te. It should be noted that the PVD method is very dependent on the growth environment, high-purity sources, and selected substrates.

Chemical vapor deposition (CVD) has also been reported to synthesis 2D Te nanoflakes. Zhang et al. developed a hydrogen-assisted CVD method and successfully synthesized ultrathin Te nanoflakes on mica substrate.[8] As illustrated in Figure 5.2(g), during the reaction proceeds, tellurium dioxide (TeO_2) was employed as Te precursor and H_2 was used as the reducing atmosphere. The as-synthesized Te nanoflakes exhibited the triangular shape, and the thickness of thinnest sample could be down to 5 nm (Figure 5.2(h)). The DFT calculations and experiments confirmed the growth mechanism origins from the formation of volatile intermediates, which increased the vapor pressure of the source and promoted the reaction.

5.2.2.3   Mechanical and Liquid-Phase Exfoliation

Top-down methods like mechanical exfoliation have played a pioneering role in discovering a series of a layered 2D monolayer, which is also successful in fabricating tellurene. Unlike other mechanical exfoliated large-area 2D materials (graphene and TMDs), Te helical chains in bulk crystal interact with each other in weak van der Waals (VDW) force, so exfoliating them will break them away laterally and vertically, resulting in the thin and narrow nanoflakes. Mechanical exfoliated ultra-thin Te nanoflakes by manually sliding a freshly cleaved facet on the silicon substrate is reported by Churchill et al., and exfoliated nanoflakes showed the widths below 100 nm and thicknesses of 1–2 nm.[9] Besides mechanical exfoliation, liquid-phase exfoliation (LPE) is another widely explored top-down method to obtain 2D Te nanoflakes. Xie et al. realized ultrathin nonlayered Te nanosheets from the bulk one by employing the LPE technique, and the obtained Te nanosheets covered a thickness ranging from 5.1 to 6.4 nm and a broad lateral size ranging from 41.5 to 177.5 nm.[10] Although the exfoliation methods provide a route to product group-VI nanosheets, it is still challenging to obtain large-area and high-quality 2D tellurene by the exfoliation technique.

In addition to the vapor deposition and exfoliation methods mentioned above, some other technologies have been developed to prepare 2D tellurene flakes. For example, Wu et al. developed a topotactic transformation strategy to obtain large-size ultrathin Te layers with clean interface,[11] where the 2D Te films were transformed from the layered MTe_2 (M=Ti, Mo W) framework by excessive lithiation, as shown in Figure 5.2(j). The resultant Te nanosheets possessed a large lateral size up to tens of micrometers and the thickness of the sample could range from a single-layer to dozens of nanometers.

5.2.3 Physical Properties

5.2.3.1 Mechanical Properties

The mechanical flexibility is a critical technical index for designing and selecting components of electronic devices. A flake structure with excellent mechanical properties would support the flexible assembly and transfer on a large scale, and especially become a good candidate in practical strain engineering. Dong et al. theoretically investigated the mechanical properties of tellurene allotropes of α-, β-, and γ-Te using DFT calculations.[12] It is found that β-Te exhibits the most obvious anisotropy in the stress-strain curve, Young's modulus, and Poisson's ratio due to intrinsic structural characteristic (Figure 5.3(a)). β-Te could endure relatively high critical strain up to 35% and 36% with the corresponding Young's modulus of 55 GPa and 27 GPa in the zigzag and armchair direction, respectively. Compared to α- and γ-Te, the higher critical strain and smaller Young's modulus indicate that β-Te possesses good toughness and superior flexibility. Moreover, under the strain smaller than 15%, easy stretching direction reverses from armchair to zigzag, which is attributed to the competition between the effective valance charge transfer along the armchair direction and puckered structure. In addition, under the strain parallel to the pucker direction of β-Te, the Poisson's ratios along the out of plane direction is negative, resulting from the intrinsic puckered structure.

5.2.3.2 Electronical Properties

Owing to the great potential of 2D Te nanoflakes in electronical and photoelectronic device applications, it is important to understand their electronic and optical properties. As emerging members of 2D materials, mono- and few-layer tellurene have received various theoretical calculations and experimental studies on electronic properties. In Figure 5.3(b), Zhu et al. firstly reported that α- and β-Te are semiconductors with the indirect and direct band gap of 0.75 and 1.47 eV, respectively,[1] while γ-Te exhibits metal characteristics. Benefitting from the nearly direct and direct band gap, both α- and β-Te also show super optical absorptions, which make tellurene promising in optoelectronic devices and photon detection. They also suggested that the carrier mobilities of α- and β-Te are much higher than monolayer $2H-MoS_2$, ranging from hundreds to thousands of $cm^2V^{-1}s^{-1}$. By applying biaxial compressive (BC) strain, the band gap of β-Te exhibits the direct-to-indirect

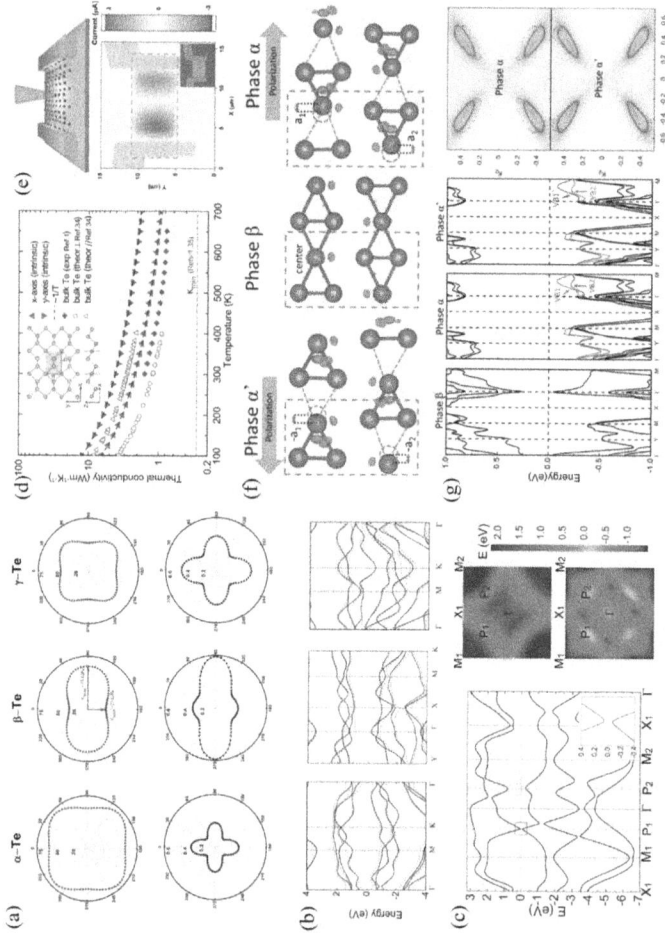

FIGURE 5.3 (a) The orientation-dependent Young's modulus and Poisson's ratio of α-, β-, and γ-Te.[1] (b) Band structures of α-, β-, and γ-Te.[12] (c) Band structures and band contour for the bottom of conduction band (upper panel) and the top of valence band (lower panel) for square Te.[3] (d) Lattice thermal conductivity of monolayer β-Te as a function of temperature.[17] (e) Schematic of the laser-induced thermoelectric current mapping of the real device.[20] (f) Differential charge density of bilayer Te and the shift of the layer-central Te atoms. (g) Band structures and spin-textures of bilayer Te phases.[21]

transition, while α-Te has a different band gap transition of firstly indirect-to-direct and then direct-to-indirect.[13] Their hole and electron effective masses could also be effectively tuned by BC strain, which in turn regulates the transport properties of tellurene. With first-principles calculations, Xian et al. predicted another stable 2D Te layered structure, named square tellurene.[3] This special chair-like buckled structure causes highly anisotropic band dispersions near the Fermi level, which can be well explained by a generalized semi-Dirac Hamiltonian. Figure 5.3(c) shows the calculated band structures including spin–orbit coupling (SOC). It is clearly shown that the Dirac-cone-like dispersions located at P_1 point to the Brillouin zone. However, in contrast to the Dirac cone in group IV 2D materials, the band dispersion around P_1 shows obvious anisotropic behavior, as shown the band contours in Figure 5.3(c). Interestingly, bulk and few-layer Te composed of helical atomic chains exhibit nearly isotropic electrical transport properties along interchain and intrachain directions. Liu et al. demonstrated that this isotropy is related to the delocalization of the lone-pair electrons, which results in similar potentials and effective masses for charge carriers along different directions.[14] And this delocalization also enhances the interchain interaction, giving rise to a fast diffusion of interstitial atoms and vacancies across Te chains. Importantly, the fast interchain and intrachain transport together enable rapid self-healing of interstitial atoms and vacancy defects at low temperature.

5.2.3.3 Thermoelectric Characteristics

Due to the low lattice thermal conductivity and superior electronic transport, bulk trigonal Te has shown great potential to be a high-performance thermoelectric material, being a preferable thermoelectric figure-of-merit.[15, 16] Furthermore, with the discovery and successful synthesis of 2D tellurene, people gradually focus on the exploration of thermoelectric properties of 2D Te nanoflakes. Here, we will highlight some representative studies. Based on the DFT calculations and the phonon Boltzmann transport equation, Gao et al. theoretically revealed the ultralow thermal properties of monolayer β-Te, which endows tellurene with great potential to be a novel thermoelectric material.[17] As shown in Figure 5.3(d), they suggested that monolayer β-Te possesses the ultralow room-temperature thermal conductivity of 4.08 and 2.16 W m^{-1} K^{-1} along the zigzag and armchair directions, respectively, which are comparable with bulk Te and are the lowest recorded values among currently reported 2D materials. This ultralow thermal conductivity could be attributed to the low-energy optical modes, the soft acoustic modes, and the strong optical–acoustic phonons scattering. Subsequently, Sharma et al. investigated the thermoelectric response of n- and p-doped 2D tellurene by combining first-principles calculations with semiclassical Boltzmann transport theory. They also proved the lowest lattice thermal conductivity of tellurene among all monoelemental 2D materials and observed a promising thermoelectric figure-of-merit ZT of 0.8 at room temperature and moderate p-doing of 1.2 ×10^{-11} cm^{-2}.[18] In addition, Lin et al. further reported the thermoelectric properties of single-layered square tellurene, which shows the three-phonon limited thermal conductivity.[19] Similar to previous investigations, the great anharmonic phonon scattering process effectively dominates and limits lattice thermal conductivity, and thermal performance along armchair direction is better than that of the zigzag direction due to the larger effective mass and Seebeck coefficient. Consequently, square tellurene exhibits the optimal ZT value of 0.79 with p-type doping and the dopant concentration of 10^{13} cm^{-2}. In experiments, Qiu et al. firstly reported the thermoelectric performance of 2D Te films (Figure 5.3(e)).[20] The room-temperature power factor and ZT value were measured to be 31.7 μW/cm K^2 and 0.63, respectively. It was also found that high work function metals can form accumulation-type metal-to-semiconductor contacts to Te films, which will efficiently collect thermoelectrically generated carriers and improve the efficiency of harvesting thermoelectric power.

5.2.3.4 Ferroelectric Properties

Two-dimensional ferroelectricity has become a very important topic in condensed state physics. Nevertheless, the most reported ferroelectric materials are compounds with different kinds of atoms. The elemental ferroelectric materials had never been reported until Wang et al. firstly introduced the concept of 2D ferroelectricity in elemental Te multilayers.[21] As shown in Figure 5.3(f), because of the interlayer interaction between lone pairs, Te multilayers exhibits the spontaneous in-plane polarization, and the estimated magnitude of polarization could reach about 1.02×10^{-10} cm^{-1} per layer. Importantly, this spontaneous in-plane polarization in bilayer Te film could be effectively maintained even above room temperature. Also, due to the strong spin–orbit coupling of Te, the hole carriers in multilayers achieve switchable valley-dependent spin-textures coupled with the direction of ferroelectricity polarization, and the textures could be effectively tuned through an external electric field (Figure 5.3(g)). Therefore, elemental Te multilayers not only expand the 2D ferroelectric family, but also provide a promising platform to explore the fascinating physical properties of 2D ferroelectric and future electronic applications. Moreover, Cai et al. demonstrated that upon the transition of strain-induced centrosymmetric phase to non-centrosymmetric phase via biaxial tensile strain, 2D ferroelectricity could also be formed in monolayer β-Te with a high polarization of about 90 μC cm^{-2}.[22]

5.2.4 DEVICE APPLICATIONS

Transistors are the fundamental building blocks of integrated circuits that are widely used in modern electronic devices. To date, some high-performance field-effect transistor (FETs) based on 2D materials have been created. For example, 2D BP-based FETs achieved a great drain current modulation up to ~10^5 and a field-effect mobility value up to ~1000 cm^2 V^{-1} s^{-1}.[23] Owing to high carrier mobility, excellent environment stability, and in-plane anisotropic properties, 2D Te nanoflakes have also been extensively explored as potential FETs. Wang et al. first demonstrated high-performance FETs based on few-layer Te films, which have a high on/off ratios on the order of ~10^6 and field-effect mobilities of ~700 cm^2 V^{-1} s^{-1} (Figure 5.4(a)).[24] Importantly, the fabricated FET devices

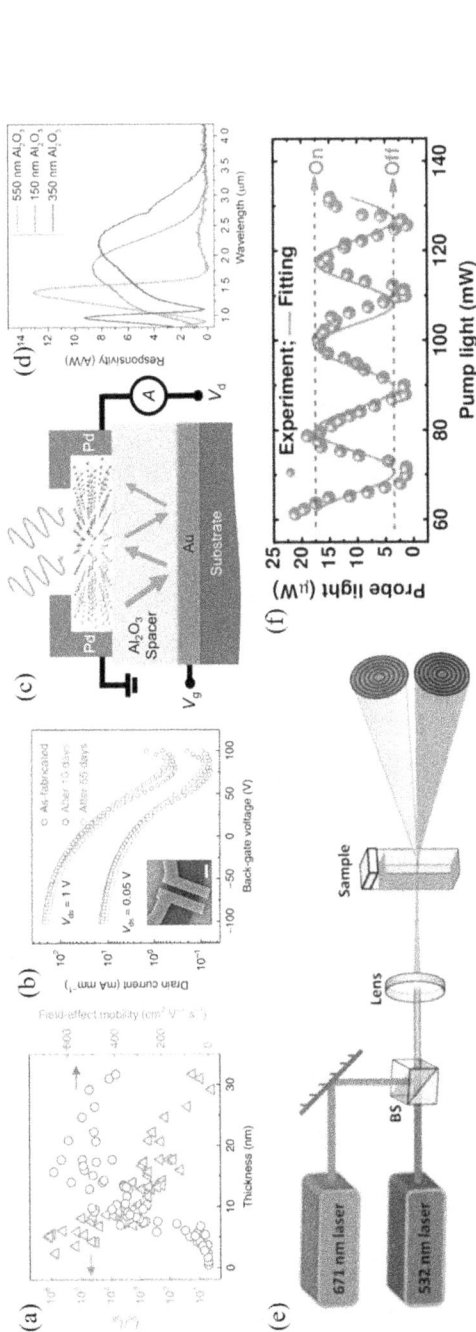

FIGURE 5.4 (a) Thickness dependent on/off ratios and field-effect mobility of 2D tellurene transistor with the thickness of 15 nm and stability measurement for up to 55 days. (b) Transfer curves of tellurene transistor with the thickness of 15 nm and stability measurement for up to 55 days.[24] (c) Schematic of the optical cavity structure used to fabricate short-wave infrared photoconductors based on quasi-2D Te nanoflakes. (d) The responsivities of Te nanoflakes fabricated on optical cavities with different Al_2O_3 thicknesses.[27] (e) Schematic of the proposed light-modulate-light system based on 2D Te Nanoflakes. (f) The achieved ON/OFF modes in Te nanoflakes-based all-optical switcher through pump light to modulate the probe light.[29]

exhibit excellent air-stability for over about two months at room temperature (Figure 5.4(b)). Zhao et al. also reported a wafer-scale p-type FETs based on thermal-evaporation synthesized Te thin films. The fabricated device with 8-nm thick film achieves an on/off current ratio of ~10^4, an effective hole mobility of ~35 cm^2 V^{-1} s^{-1}, and subthreshold swing of 108 mV dec^{-1}.[25] In addition, high-performance electrolyte-gated transistors (EGTs) have also been reported and the insulator–metal transition of EGTs can be achieved by tuning gate voltage, then achieving mobility up to ~500 cm^2 V^{-1} s^{-1}.[26]

5.2.4.1 Photodetector

Photodetectors are another potential device application for 2D Te nanoflakes. Amani et al. reported short-wave infrared photodetectors based on solution-synthesized and air-stable quasi-2D Te flakes. Despite their indirect band gap, the fabricated devices still exhibit high-performance photoresponse by employing Au/Al$_2$O$_3$ optical cavity substrates to further increase the absorption (Figure 5.4(c)).[27] As shown in Figure 5.4d, by adjusting the thickness of Al$_2$O$_3$ spacer, the peak responsivity of Te gated photodetector can capture the short-wave infrared band from 1.4 µm to 2.4 µm with a cutoff wavelength of 3.4 µm. Subsequently, Shen et al. demonstrated the use of hydrothermal-synthesized Te nanoflakes as an ultrasensitive and broadband photodetector, which exhibits a high anisotropic behavior and a large bandwidth of 37 MHz at the communication wavelength of 1.55 µm.[28] Te photodetector presents the peak extrinsic responsivity of 18.9 mA W^{-1}, 19.2 mA W^{-1}, and 383 A W^{-1} at 3.39 µm, 1.55 µm, and 520 nm light wavelength, respectively. Owing to the photogating effect, the fabricated device achieves high gains up to 1.9×10^3 and 3.15×10^4 at 520 nm and 3.39 µm wavelength, respectively. Furthermore, Deckoff-Jones et al. proposed and designed a waveguide-integration of mid-infrared tellurene optoelectronic devices, which can achieve an extremely low noise photodetector at room temperature.

In addition to FETs and photodetectors, 2D Te has become a new nonlinear optical material and shows great potential in nonlinear photonic application. For example, Wang et al. designed a light-modulate-light system based on 2D Te for all-optical modulation applications, and the schematic of the system by means of spatial cross-phase modulation is shown in Figure 5.4(e).[29] The proposed all-optical modulator can realize the "ON" and "OFF" modes by taking advantage of stronger-power pimp light to modulate other weaker-power probe light (Figure 5.4(f)). The successful application of 2D Te nanoflakes in an all-optical modulator leads to an important step for the 2D group-VI materials toward nonlinear optical devices.

5.3 SELENENE

5.3.1 Crystal Structure of Selenene Allotropes

The successful discovery of tellurene further stimulates the interest in searching for 2D elemental Se materials (selenene). In the previous study as Zhu et al. demonstrated, the underlying formation mechanism of tellurene is inherently rooted in the multivalent nature, and they also further suggested the importance of multivalency in the 2D structure of Se and predicted a 1T-MoS$_2$-like monolayer (Figure 5.5(a)). Subsequently, various crystals of Se monolayers are theoretically proposed and examined. Figure 5.5 shows three possible monolayer structures of Se via theoretical prediction: t-Se, c-Se, and s-Se.[2] The t-Se phase exhibits a similar 1T-MoS$_2$-like sandwich structure to α-Te (Figure 5.5(a)). There are two distinct types of Se atoms with different coordination numbers (Nc). The central Se atom in the middle layer has Nc = 6, while the Se atom in the lower and upper layer has Nc = 3, which is different from most single elemental 2D materials. The c-Se phase consists of a series of weakly interacted helical chains, which are parallel-arranged on a plane as shown in Figure 5.5(b). In Figure 5.5(c), the monoclinic s-Se has a buckled square configuration containing two atoms per unit cell, in which one atom is located at the corner of the square lattice and the other one in the center is titled toward the corner direction. Among these three crystal phases, c-Se with the lowest formation energy is the most stable phase, and Se nanosheets with c-Se phase have been successfully synthesized in most experiments.[30, 31]

5.3.2 PREPARATION METHODS

Physical vapor deposition (PVD)꞉ vapor deposition has also been extended to 2D Se growth. Using the PVD method, Qin et al. demonstrated the preparation of large-size, high-quality 2D selenium nanosheets.[30] Figure 5.6(a) shows the typical growth setup, in which Se powder was placed in a quartz glass tube as the precursor. The as-grown Se nanosheets on a Si substrate are uniformly distributed on the surface of polycrystalline Se films. The SEM image in Figure 5.6(b) clearly shows that most of the nanosheets exhibit a saw-like structure with the average

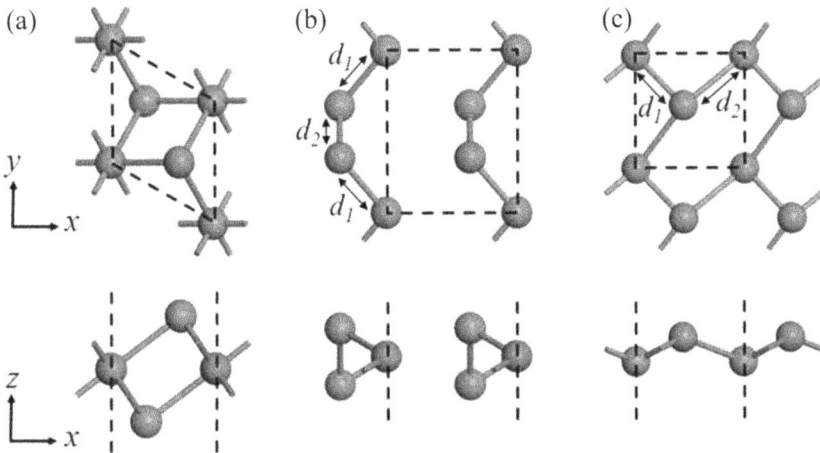

FIGURE 5.5 Top and side views of three optimized structures of 2D selenium: (a) t-Se, (b) c-Se, and (c) s-Se.

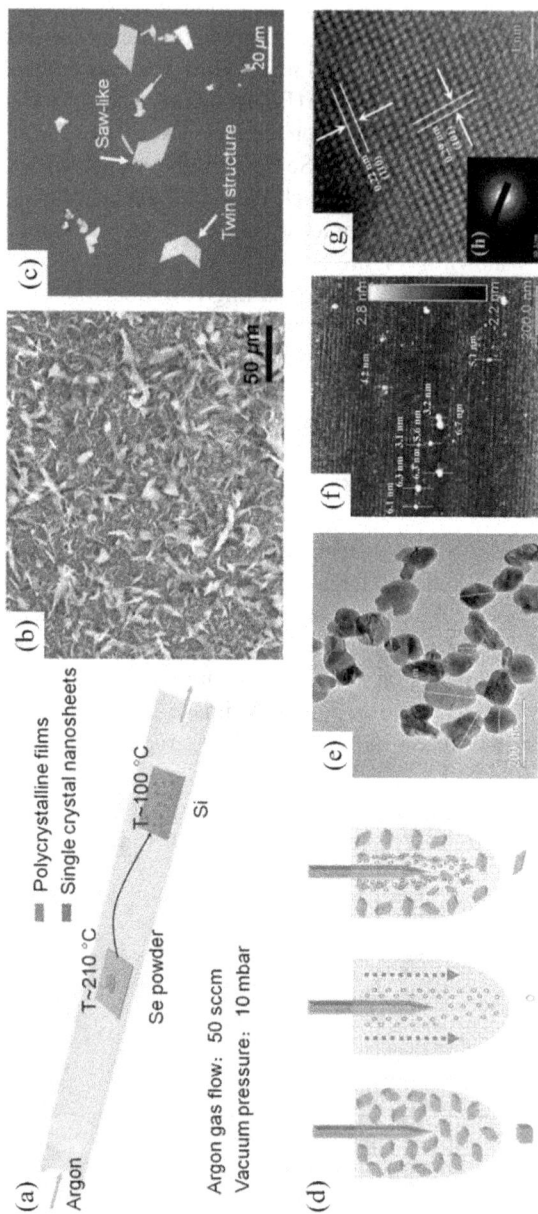

FIGURE 5.6 (a) Schematic diagram of the physical vapor deposition method. (b) Low-magnification SEM image of Se nanosheets on Si (111) substrate. (c) Optical microscopy image of Se samples on the SiO₂/Si substrate.³⁰ (d) Schematic of a probe-sonication liquid-phase exfoliation method illustrating the exfoliation process of 2D nonlayered Se nanosheets from bulk Se powder. (e) TEM images showing the lateral size. (f) AFM images showing the thickness of the Se nanosheets. (g) High-resolution TEM images describing the lattice fringes of as-prepared Se nanosheets.³¹

width of about 8 μm and the maximum length up to 50 μm. Consistent with the SEM image, optical microscopy image in Figure 5.6(c) further identifies the surface morphology of the Se nanosheets, in which most of them have an irregular quadrangular shape with zigzag edges, and the others exhibit twin structures with mirror symmetry. The growth of Se nanosheets could be well explained by the vapor–solid mechanism. It is found that the high chemical activity at the vertexes and ridges of 1D nanorods plays a critical role for nanosheet growth. As the reaction continues, the growth of the Se nanostructure is gradually driven by effective activation sites count and surface energy. In the initial crystallization, surface energy difference dominates the growth of the 1D nanorod. When the end cavity of nanorod is saturated, Se_2 molecules would impinge on the surfaces of the nanorod and diffuse from site to site, finally leading to the production of 2D nanosheets.

For layered materials, liquid-phase exfoliation is an effective method to obtain 2D layered nanosheets. Interestingly, even though bulk Se power is nonlayered materials, 2D Se nanosheets can also be exfoliated from the nonlayered bulk Se via liquid-phase exfoliation. Xing et al. reported, for the first time, the fabrication of ultrathin nonlayered 2D Se through a probe-sonication liquid-phase exfoliation method, as shown in Figure 5.6(d).[31] The as-prepared 2D Se nanosheets have the lateral sizes from 40 to 124 nm (Figure 5.6(e)) and an average thickness of 3–6 nm (Figure 5.6(f)). The high-resolution TEM image and the selected-area electron diffraction (SAED) image in Figure 5.6(g) confirmed a typical trigonal phase similar to bulk Se, indicating the highly crystalline features of as-prepared 2D nanosheets during the exfoliation procedure. It is found that the successful fabrication of 2D Se nanosheets from nonlayered bulk Se may originate from two kinds of anisotropy including bulk Se itself and the probe sonication in the vertical direction.

5.3.3 Physical Properties

Figure 5.7(a) shows the band structures without/with SOC effect of t-, c-, and s-Se, respectively, obtained within the PBE functional.[2, 32] It is found that both t-Se and c-Se are semiconductors and have indirect band gaps of 1.04 and 2.57 eV within a more accurate HSE06 hybrid scheme, respectively. Furthermore, as shown in Figure 5.7(b), the t-Se monolayer ($α$-Se in Ref. 32) is predicated to possess high carrier mobility with 6.97×10^3 and 9.48×10^3 cm² V^{-1} s^{-1} for electron and hole, respectively, which are higher than those (2.09×10^3 and 1.76×10^3 cm² V^{-1} s^{-1} for electron and hole) of $α$-Te.[32] The high mobilities for both electron and hole carriers endow t-Se with bipolar conductivity. On the other hand, the c-Se monolayer ($β$-Se in Ref. 32) exhibits poor anisotropic carrier mobility for both electron and hole. The calculated highest hole mobility of c-Se can reach 571 cm² V^{-1} s^{-1} comparable with other conventional 2D materials. Interestingly, for s-Se, there exist Dirac-cone-like dispersions in the band structure, and these band dispersions are highly anisotropic, which are different from the Dirac cones in 2D group-IV materials.[2, 3] It is also found that the special square unit cell gives rise

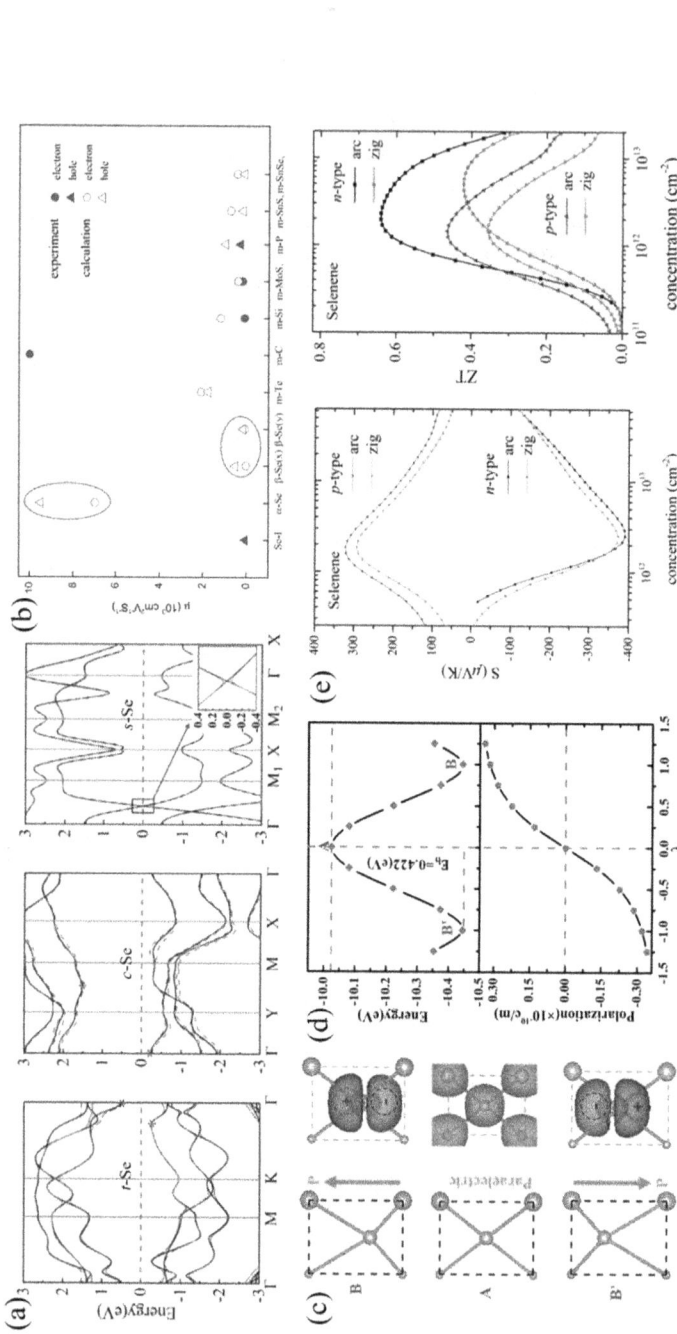

FIGURE 5.7 (a) Band structures of *t*-Se, *c*-Se, and *s*-Se without/with SOC effect indicated by the solid and dashed lines, respectively. (b) Comparison of the carrier mobilities of selenene with some conventional 2D materials. (c) Top view of the unit cell of *c*-Se monolayer in three different states and the corresponding charge density difference. (d) The variation of total energy (top) and polarization (bottom) as a function of percentage of the change in atomic position between centrosymmetric and polar structures.[32] (e) Seebeck coefficient and ZT value along the zigzag and the armchair direction at 300 K of *c*-Se monolayer.[19]

to a highly anisotropic transmission characteristic of s-Se monolayer, which has a larger carrier mobility of 93.2×10^3 cm^2 V^{-1} s^{-1} for electron and 42.2×10^3 cm^2 V^{-1} s^{-1} for hole than those of square tellurene.[19]

Owing to the breaking of central inversion symmetry makes, c-Se exhibits a considerable in-plane spontaneous polarization of 13.46 μCcm^{-2}, which is comparable to that of conventional perovskite ferroelectrics, such as bulk BaTiO$_3$ (25 μCcm^{-2}).[32] Unlike 2D compounds with a positive and negative ion center, total polarization of c-Se monolayer is only derived from the electronically induced dipole moment, and spontaneous electric polarization origins from the divided equivalent positive-negative ion pair due to charge density distortion (Figure 5.7(c)). Importantly, in the top of Figure 5.7(d), the reversing barrier of polarization of 422 meV per unit cell indicated that 2D elemental ferroelectricity in c-Se can be well stabilized at room temperature. It is also found that the variation of polarization with distortion mode is nonlinear and exhibits the trend of hyperbolic tangent.

Similar to square tellurene, c-Se also exhibits good thermoelectric property and has high three-phonon limited thermal conducticity.[19] As shown in Figure 5.7(e), the Seebeck coefficient of c-Se has a maximum Seebeck value of 400 and 300 μVK^{-1} for n- and p-type doping in the dopant range 10^{12}–10^{14} cm^{-2}, indicating the obvious bipolar effect. The special square unit cell of c-Se monolayer leads to a highly anisotropic Seebeck structure and the heavier effective mass in the armchair direction gives rise to larger Seebeck coefficient than that in the zigzag direction for whatever n- and p-type doping. Finally, c-Se monolayer exhibits a good thermoelectric property by n-type doping, and maximum ZT value can reach 0.64 at the moderate dopant concentration of near 2×10^{12} cm^{-2} (Figure 5.7(f)).

5.3.4 DEVICE APPLICATIONS

Back-gating field-effect transistors (FETs) based on 2D Se nanosheet have also been demonstrated, as shown in the inset of Figure 5.8(a). Qin et al. reported that Se nanosheets FETs with Ni/Au as metal contacts exhibit typical p-type transport behavior with a high current on/off ratio over 10^6 and a relatively low on-state current density around 20 mA/mm at V$_{ds}$ = 3 V (Figure 5.8(a)).[30] The p-type conductivity of Se FETs mainly arises from the presence of hydroxyl and hydrogen terminations on the nanosheets surface. Furthermore, they fabricated Se nanosheet phototransistors and examined the optoelectronic performance, as shown in the schematic diagram in Figure 5.8(c). The device exhibits a pronounced photoresponsivity of 263 A/W even at a very low illumination power down to 0.21 mW cm^{-2}. It is also found that the responsivity is linearly proportional to the illumination power, indicating that photoexcited carriers mainly determine the photocurrent (Figure 5.8(d)). Subsequently, Fan et al. also demonstrated a 2D Se-based photodetector by the photoelectrochemical method.[33] As shown in Figure 5.8(e), the fabricated ultra-thin photodetector can exhibit obvious photoelectronic response at bias voltage of 0 V, suggesting the great potential

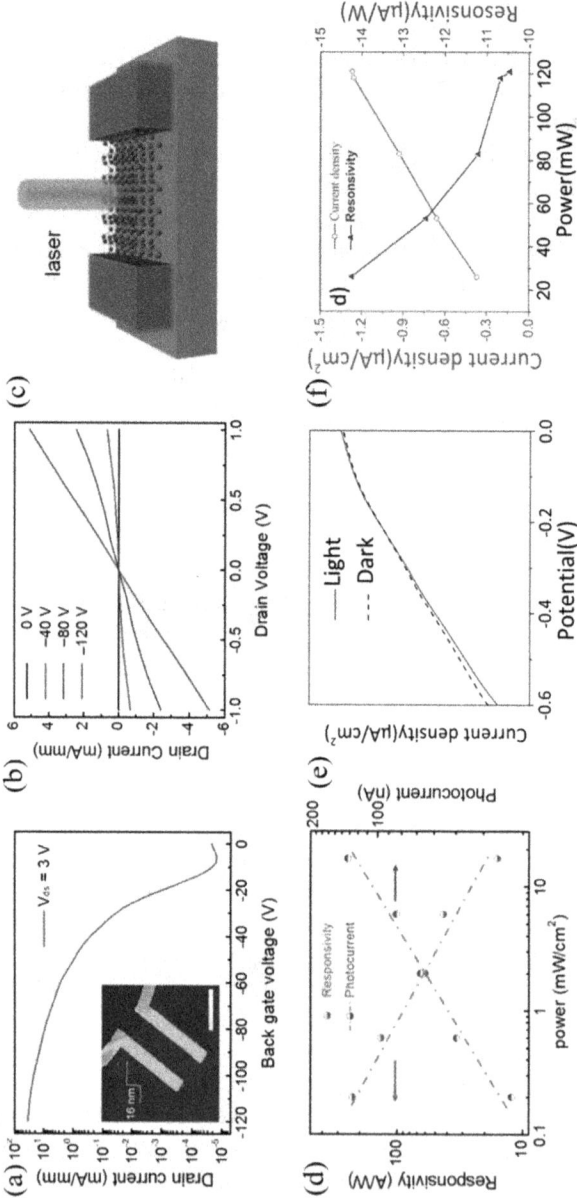

FIGURE 5.8 (a) Transfer performance of Se nanosheet field effect transistor device with a thickness of 16 nm and the AFM height profile of the device. (b) Output performance of the same Se nanosheet transistor. (c) Schematic diagram of a Se nanosheet-based phototransistor. (d) Measured responsivity and photocurrent as a function of laser illumination power.[30] (e) Linear sweep voltammetry test of Se nanosheet-based photodetector in 0.1 M KOH. (f) Photocurrent density and responsivity of Se nanosheet-based photodetector under various bias potential in 0.1 M KOH.[31]

as self-powered photodetector. At the bias voltage of -0.6 V and the irradiation intensity of 121 mW cm^{-2}, the photodetector will achieve high photocurrent density of 1.28 mW cm^{-2} and photoresponsivity of 10.45 µA W^{-1} (Figure 5.8(f)). In addition, long-term photoelectronic measurements demonstrate that 2D Se-based photodetector possesses a good cycle stability. Furthermore, Xing et al. demonstrated a 2D Se nanosheet-based optical modulation device that allows for excellent ultrashort pulse generation of an optical communication band.[31]

5.4   SUMMARY AND PERSPECTIVES

In summary, we have summarized the crystal structures, synthesis methods, basic physical properties, and promising device applications of emerging 2D group VIA materials: tellurene and selenene. In addition to excellent electronic properties and great potential in nanoelectronics as mentioned in this chapter, present investigations on 2D Te and Se materials also open up some significant opportunities and challenges for the further development of 2D group VIA materials.

Firstly, the synthesis method of 2D Te and Se is one of the most challenging problems. Developing effective methods for achieving massive production of high-quality and large-dimension 2D Te/Se-based materials is still in progress and is now a great bottleneck for the large-scale production and performance of the devices. In addition, realizing controllable synthesis of nanosheets with desired morphology and thickness has also become a critical issue for device applications. Secondly, although recent studies have shown the promising application of 2D Te/Se nanosheets in FETs and photodetectors, the behind carrier transport and dynamics mechanisms in these devices are still ambiguous. And apart from the aforementioned applications, further extending the applications of 2D Te/Se materials to other novel fields is also very significant for future development, such as biomedicine, highly integrated chips, and flexible electronic devices, etc.

For the crystal structures themselves of tellurene and selenene, there are still some differences between theoretical predictions and experimental synthesis results. Theoretical predictions have demonstrated several stable crystal allotropes, while most of the experimentally reported 2D think films only exhibits nonlayered structures with helical chains. At the same time, the allotropes' growth and phase transition mechanisms are not clearly understood, waiting to be disclosed. Therefore, much more theoretical and experimental works are needed to investigate new preparation schemes to achieve controlled growth of high-quality samples and prospective nanodevice applications of 2D Te/Se materials, and to further discover the related physical mechanisms about carrier dynamics, structural growth, and phase transition, etc.

REFERENCES

1. Zhu, Z. et al., Multivalency-driven formation of te-based monolayer materials: a combined first-principles and experimental study. *Physical Review Letters*, 119, 10, 2017.

2. Liu, C. et al., 2D selenium allotropes from first principles and swarm intelligence. *Journal of Physics: Condensed Matter*, 31, 23, 2019.
3. Xian, L., Pérez Paz, A., Bianco, E., Ajayan, P. M. & Rubio, A., Square selenene and tellurene: novel group VI elemental 2D materials with nontrivial topological properties. *2D Materials*, 4, 4, 2017.
4. Chen, J. et al., Ultrathin beta-tellurium layers grown on highly oriented pyrolytic graphite by molecular-beam epitaxy. *Nanoscale*, 9, 41, 2017.
5. Huang, X. et al., Epitaxial growth and band structure of Te film on graphene. *Nano Letters*, 17, 8, 2017.
6. Yang, S. et al., Highly crystalline synthesis of tellurene sheets on two-dimensional surfaces: control over helical chain direction of tellurene. *Physical Review Materials*, 2, 10, 2018.
7. Apte, A. et al., Polytypism in ultrathin tellurium. *2D Materials*, 6, 1, 2018.
8. Zhang, X. et al., Hydrogen-assisted growth of ultrathin te flakes with giant gate-dependent photoresponse. *Advanced Functional Materials*, 29, 49, 2019.
9. Churchill, H. O. H. et al., Toward single atom chains with exfoliated tellurium. *Nanoscale Research Letters*, 12, 1, 2017.
10. Xie, Z. et al., Ultrathin 2D nonlayered tellurium nanosheets: facile liquid-phase exfoliation, characterization, and photoresponse with high performance and enhanced stability. *Advanced Functional Materials*, 28, 16, 2018.
11. Peng, J. et al., Two-dimensional tellurium nanosheets exhibiting an anomalous switchable photoresponse with thickness dependence. *Angewandte Chemie*, 57, 41, 2018.
12. Dong, Y. et al., Study on the strain-induced mechanical property modulations in monolayer Tellurene. *Journal of Applied Physics*, 125, 6, 2019.
13. Xue, X.-X. et al., Strain tuning of electronic properties of various dimension elemental tellurium with broken screw symmetry. *Journal of Physics: Condensed Matter*, 30, 12, 2018.
14. Liu, Y., Wu, W. & Goddard, W. A., 3rd, tellurium: fast electrical and atomic transport along the weak interaction direction. *Journal of the American Chemical Society*, 140, 2, 2018.
15. Lin, S. et al., Tellurium as a high-performance elemental thermoelectric. *Nature Communications*, 7, 2016.
16. Peng, H., Kioussis, N. & Snyder, G. J., Elemental tellurium as a chiralp-type thermoelectric material. *Physical Review B*, 89, 19, 2014.
17. Gao, Z., Tao, F. & Ren, J., Unusually low thermal conductivity of atomically thin 2D tellurium. *Nanoscale*, 10, 27, 2018.
18. Sharma, S., Singh, N. & Schwingenschlögl, U., Two-dimensional tellurene as excellent thermoelectric material. *ACS Applied Energy Materials*, 1, 5, 2018.
19. Lin, C., Cheng, W., Chai, G. & Zhang, H., Thermoelectric properties of two-dimensional selenene and tellurene from group-VI elements. *Physical Chemistry Chemical Physics*, 20, 37, 2018.
20. Qiu, G. et al., Thermoelectric performance of 2D tellurium with accumulation contacts. *Nano Letters*, 19, 3, 2019.
21. Wang, Y. et al., Two-dimensional ferroelectricity and switchable spin-textures in ultra-thin elemental Te multilayers. *Materials Horizons*, 5, 3, 2018.
22. Cai, X., Ren, Y., Wu, M., Xu, D. & Luo, X., Strain-induced phase transition and giant piezoelectricity in monolayer tellurene. *Nanoscale*, 12, 1, 2020.
23. Li, L. et al., Black phosphorus field-effect transistors. *Nature Nanotechnology*, 9, 5, 2014.
24. Wang, Y. et al., Field-effect transistors made from solution-grown two-dimensional tellurene. *Nature Electronics*, 1, 4, 2018.

25. Zhao, C. et al., Evaporated tellurium thin films for p-type field-effect transistors and circuits. *Nature Nanotechnology*, 15, 1, 2020.
26. Ren, X. et al., Gate-tuned insulator-metal transition in electrolyte-gated transistors based on tellurene. *Nano Letters*, 19, 7, 2019.
27. Amani, M. et al., Solution-synthesized high-mobility tellurium nanoflakes for shortwave infrared photodetectors. *ACS Nano*, 12, 7, 2018.
28. Shen, C. et al., Tellurene photodetector with high gain and wide bandwidth. *ACS Nano*, 14, 1, 2020.
29. Wu, L. et al., 2D Tellurium based high-performance all-optical nonlinear photonic devices. *Advanced Functional Materials*, 29, 4, 2019.
30. Qin, J. et al., Controlled growth of a large-size 2D selenium nanosheet and its electronic and optoelectronic applications. *ACS Nano*, 11, 10, 2017.
31. Xing, C. et al., 2D nonlayered selenium nanosheets: facile synthesis, photoluminescence, and ultrafast photonics. *Advanced Optical Materials*, 5, 24, 2017.
32. Wang, D. et al., High bipolar conductivity and robust in-plane spontaneous electric polarization in selenene. *Advanced Electronic Materials*, 5, 2, 2019.
33. Fan, T., Xie, Z., Huang, W., Li, Z. & Zhang, H., Two-dimensional non-layered selenium nanoflakes: facile fabrications and applications for self-powered photo-detector. *Nanotechnology*, 30, 11, 2019.

6 Heterostructures Based on 2D Xenes Materials

Gencai Guo and Siwei Luo

CONTENTS

6.1 INTRODUCTION

Research on graphene and other 2D monoelemental materials is intense and has been one of the leading topics for many years. Beyond that, isolate atomic planes can also be reassembled into designer heterostructures made layer by layer in a precisely chosen sequence. It could be van der Waals heterostructure or lateral heterostructure according to the arrangement of heterostructure.

The field of 2D heterostructures has gained increasing interest in recent years with the increasing types of 2D materials and the progress in synthesis and processing. Heterostructures could not only overcome the inherent limitations of individual material, but also bring unusual properties and new phenomena.

DOI: 10.1201/9781003207122-6

Constructing heterostructures based on 2D Xenes materials not only improves the property but also boosts the development of 2D Xenes materials. Firstly, heterostructures constructed by 2D Xenes materials can be combined by van der Waals forces without requisition of lattice matching, which makes it easy to obtain. Secondly, some of the group VA materials, for example, black phosphorene (BP), are unstable under ambient conditions. Fortunately, it is found that the construction of 2D heterostructure can compensate this disadvantage. By making BP/h-BN and BP/ZnO heterostructure, the environmental stability has improved effectively.[1, 2] Thirdly, another important feature that hinders the application of 2D Xenes is the band gap, especially the band gap (0 eV) in graphene. Heterostructures can open the band gaps in 2D materials, resulting in wider application in the field of microelectronics. All the above factors have indicated that constructing heterostructures based on 2D Xenes materials is an effectively way for the development of 2D Xenes.

6.2 SYNTHESIS METHODS OF 2D XENE HETEROSTRUCTURES

The basic principle of the fabrication of 2D Xene heterostructures is simply stacking isolated 2D crystals together. In 2010, Dean et al. synthesized 2D layer materials through mechanical stacking method.[3] They used mechanical exfoliation to strip the h-BN onto the substrate and the graphene onto the PMMA. The graphene is then carefully aligned with the h-BN and touched with each other (Figure 6.1a). By repeating this process, any number of layers of different materials can be stacked, thus increasing the complexity of the heterostructures (Figure 6.1b).[4] However, this approach is likely to introduce some pollution into the interface of the layers, thus weakening the performance of the heterostructure. Therefore, dry transfer methods have also been developed. In the dry transfer process, the first exfoliated flakes are repeatedly peeled to obtain the final thin layer structure for the construction of heterostructure (Figure 6.1c).[5] This method can obtain a clean

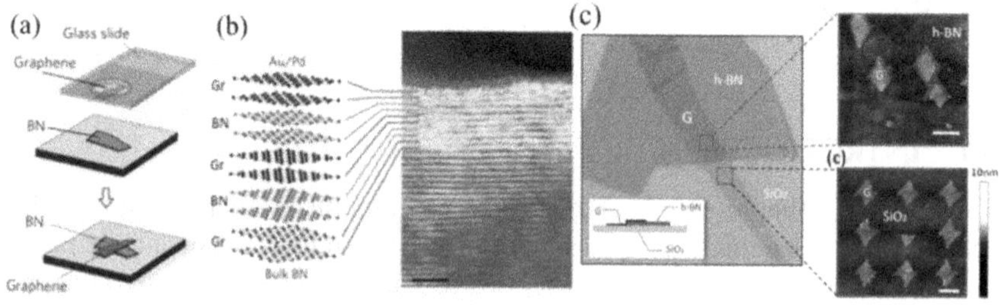

FIGURE 6.1   (a) Schematic of the alignment and stacking of graphene flake on h-BN. Reproduced with permission.[3] Copyright 2010, *Springer Nature*. (b) Bright-field cross sectional STEM of a stack of graphene and hBN bilayers with the layer sequence schematically shown to the left. Reproduced with permission.[4] Copyright 2012, *Springer Nature*. (c) AFM image. Reproduced with permission.[5] Copyright 2016, APS.

heterostructure interface. For materials that are unstable in the air, the process can be carried out in an inert gas atmosphere.

Using layers exfoliated from the bulk material, the heterostructure can be obtained quickly without developing new methods of material growth. Moreover, the interface of heterostructure can be stacked directly without lattice matching. In addition, the stacking angle between different layers can be controlled with high precision, thus the characteristics of the heterostructures can be regulated. Therefore, these methods are widely used in basic research to generate a large number of different heterostructures, such as graphene/MoS_2, graphene/h-BN, and so on.[6–8]

From the point of view of scientific research, the mechanical stacking method is very convenient to prepare a heterostructure, but for industrial application, this method is not suitable. In order to produce heterostructure materials in large quantities, it is necessary to develop new experimental preparation methods. The bottom-up growth method was used to synthesize graphene heterostructures, including graphene/MoS_2, graphene/WS_2, graphene/$MoSe_2$, graphene/WSe_2, graphene/NbS_2, graphene/$NbSe_2$, and other heterostructures.[9–24]

In addition to the method based on mechanical stacking, several methods have been developed to synthesize heterostructures. Liu et al. reported a two-step synthesis of h-BN/graphene heterostructure.[9] First, CVD was used at 950°C and graphene was grown on copper foil using hexane as carbon feedstock. Then, h-BN was grown on graphene using a second CVD with ammonia borane as the precursor. Recent studies have shown that multilayer graphene growth on h-BN flakes using the CVD method was performed at 1000°C using CH_4 as the carbon feedstock. The mechanism of graphene growth on h-BN is not completely clear yet, but it is likely that the influence on copper is involved. In addition, the growth of molecular beam epitaxy of graphene on exfoliated h-BN has also been investigated, although the quality of graphene is not as good as that of the CVD method.[25] Roth et al. demonstrated that using h-BN/Cu (111) and h-BN/Rh (111) as growth substrates graphene could be grown with 3-pentanone at 830°C and 880°C, respectively.[26, 27]

6.3 VAN DER WAALS HETEROSTRUCTURE

Considering the combination mode of heterostructure, two types of heterostructure could be classified, van der waals heterostructure and lateral heterostructure. The van der Waals heterostructure usually exhibits anisotropic characteristics, with covalent bonds' interaction within the layer and van der Waals interactions between the layers. However, the lateral heterostructure is formed by combining two materials into a planar structure, and the interaction inside is covalent bond interaction, as shown in Figure 6.2.

Here are a few questions worth thinking about. What new properties can be obtained from heterostructures? Will these new properties exceed the original 2D material? In this chapter, VDW heterostructure is introduced.

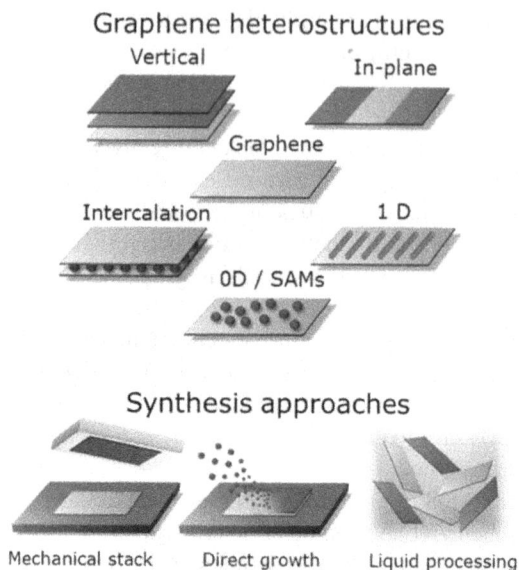

FIGURE 6.2   Schematic overview of van der Waals heterostructure and lateral hetero-structure, and the methods employed to fabricate them. Reproduced with permission.[28] Copyright 2017, Royal Society of Chemistry.

6.3.1   GROUP IVA

6.3.1.1   Graphene-Based Heterostructures

The novel physical and mechanical properties of graphene hint at the great potential of it in 2D heterostructures.[28] In the meanwhile, considering the wide range of 2D materials and their special properties, combining graphene with that of 2D materials would be very interesting and promising. Heterostructure could combine the properties of individual materials and produce many interesting electronic and optoelectronic properties. In a graphene heterostructure, graphene could provide high conductivity, carrier mobility, transparency, and mechanical flexibility.

There are many applications for graphene-based heterostructures, the most interesting of which are in the field of optoelectronics. These devices can be used to control, detect, or generate light. Graphene has great potential for photoelectric applications due to its ability to interact with a wide range of wavelengths of light, as well as its high response speed and carrier mobility and flexibility. However, its special band structure (without band gap) makes it difficult to be widely used in optoelectronic devices. On the other hand, most single-layer TMDs are direct bandgap semiconductors and exhibit large optical absorptions. Therefore, it is possible to use the interaction between graphene and TMDs to regulate their electronic and optoelectronic properties by constructing graphene/TMDs heterostructures.[29]

Lithium-ion batteries (LIBs) are a kind of energy storage device which can be periodically charged and discharged. Due to the advantages of high energy density, sustainable charge and discharge, high safety performance, environmentally friendly, LIBs have been widely used in various fields such as energy storage devices, intelligent electronic devices, gas sensors, and others. The anode used in commercial LIBs is mainly graphite, but its further application is limited by its low theoretical capacity (372mA h g^{-1}) and weak lithium ion adsorption energy. The 2D heterostructure materials have attracted extensive attention because of their special structural characteristics and the ability to integrate the advantages of a variety of isolated materials. By creating these heterostructure materials, the required high conductivity, high capacity, and long-term cycling stability can be achieved. The graphene/MoS$_2$ heterostructure exhibits high conductivity and high capacity (1675 mA h g^{-1}).[30–33]

Liu et al. investigated the potential of graphene/phosphorene heterostructure as anode material of LIBs through first-principles calculations.[34] The calculated results show that compared to pristine phosphorene or graphene, the binding energy of Li is greatly improved in graphene/phosphorene heterostructure and without the sacrifice of Li mobility. The large Li adsorption energy and fast migration ability of graphene/phosphorene heterostructure are the origin of the interfacial synergy effect. Importantly, the graphene/phosphorene heterostructure exhibits excellent stiffness (C$_{ac}$=350 N/m, C$_{zz}$=464 N/m), which is beneficial for the improvement of the cycle performance of the battery. The high capacity, good electronic/ionic conductivity, and excellent stiffness of graphene/phosphorene heterostructure indicate the great potential of it as anode material in LIBs.

6.3.1.2 Silicene-Based Heterostructure

Silicene based heterostructure are predicted to have interesting physical properties. However, the experimental preparation is very difficult because silicene is unstable in the air. Gao et al. successfully fabricated silicene/graphene heterostructure by silicon intercalation.[35] The heterostructures exhibit good air stability and maintain good integrity after a long time of air exposure.

Silicene is considered to be a good material for the anode of LIBs and has the advantages of high capacity and low Li migration energy barrier. But silicene is generally unstable and needs to be stabilized by forming heterostructures with other materials. Xu et al. proposed to construct a silicene/graphene heterostructure to protect the silicene and further investigated the potential of this heterostructure for lithium/sodium ion batteries, as shown in Figure 6.3.[36] The calculated results show that the silicene/graphene heterostructures not only preserve the high lithium/sodium storage capacity of silicene and the low migration energy barrier of lithium/sodium, but also provide a larger binding energy of lithium/sodium, which can effectively inhibit the formation of lithium dendrites. The silicene/graphene heterostructures are all metallic before and after lithium/sodium intercalation and are very beneficial to electron transport. Furthermore, the stiffness of the silicene/graphene heterostructure is greater than that of the

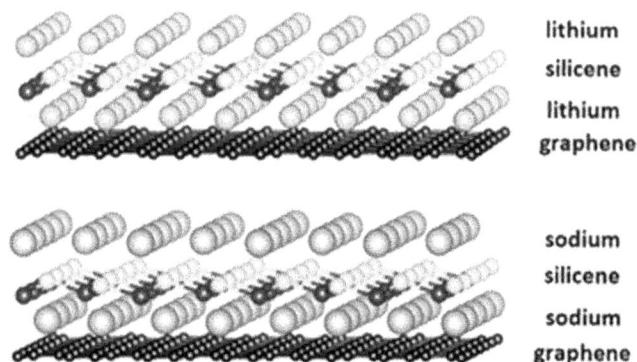

FIGURE 6.3 The lithium/sodium storage mechanism of silicene/graphene heterostructures. Reproduced with permission.[36] Copyright 2016, Royal Society of Chemistry.

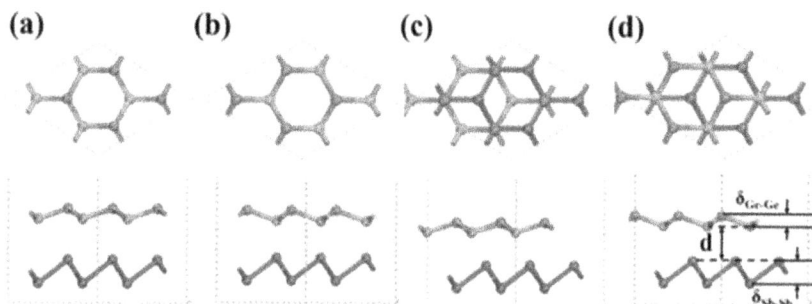

FIGURE 6.4 Top and side view of the atomic model of the germanene/antimonene heterostructure. (a) Pattern AAI, (b) Pattern AAII, (c) Pattern ABI, (d) Pattern ABII. Reproduced with permission.[37] Copyright 2016, Royal Society of Chemistry.

original silicene or graphene and contributes to the improvement of the cycling performance of the electrode.

6.3.1.3 Germanene-Based Heterostructure

Chen et al. investigated the electronic and optical properties of germanene/antimonene heterostructures through first-principles calculations.[37] The structure of germanene/antimonene heterostructure are shown in Figure 6.4. The calculated results show that the germanene and antimonene monolayers are bound to each other via orbital hybridization and can open the band gap. Our results show that the AAII structure has a direct band gap (391 meV), while the other three structures have an indirect band gap, which can be tunable between 37 and 171 meV. The band gap of germanene/antimonene heterostructures can be adjusted in a wide range by changing the direction and intensity of the external electric field. Furthermore, germanene/antimonene heterostructures exhibit more obvious optical conductivity.

Zhao et al. investigated the transition from Schottky to Ohmic contacts in janus germanene/MoSSe heterostructures.[38] The calculated results show that Ge/SMoSe forms n-type Schottky contact, and Ge/SeMoS forms p-type Schottky contact. A transition from Schottky to Ohmic behavior occurs when tensile strain is present (Ge/SMoSe: 4%, Ge/SeMoS: 8%). Furthermore, the results indicate that increasing the thickness of MoSSe could lead to an Ohmic contact. These results indicate that germanium is an ideal electrode for contacting 2D Janus Mosse in electronic devices.

6.3.1.4   Stanene-Based Heterostructure

Recently, Maliha et al. have studied the thermal conductivity of the stanene/silicene heterostructure and bilayer stanene nanostructures to characterize their thermal transfer phenomena.[39] The calculated results revealed the good thermal stability within the temperature range of 100–600 K of these structures. The thermal conductivities at room temperature of original stanene/silicene heterostructure and stanene bilayer are estimated to be 3.63 ± 0.27 W m^{-1} K^{-1} and 1.31 ± 0.34 W m^{-1} K^{-1}. When the temperature changes from 100K to 600 K, the thermal conductivity of the studied bilayer nanoribbons decreases with increasing temperature.

Liang et al. investigate the geometric and electronic properties of stanene/MoS$_2$ heterostructure by first-principles calculations; the structures are shown in Figure 6.5.[40] The calculated result shows that the stanene can absorb on the MoS$_2$ forming stanene/MoS$_2$ heterostructures. The Dirac point of stanene is still preserved on stanene/MoS$_2$ heterostructure. Furthermore, the band gap of stanene/MoS$_2$ heterostructure is 67 meV and can be effectively modulated through an external strain or a perpendicular electric field. These results are helpful for the research of tunability of the electronic properties in stanine-based heterostructures.

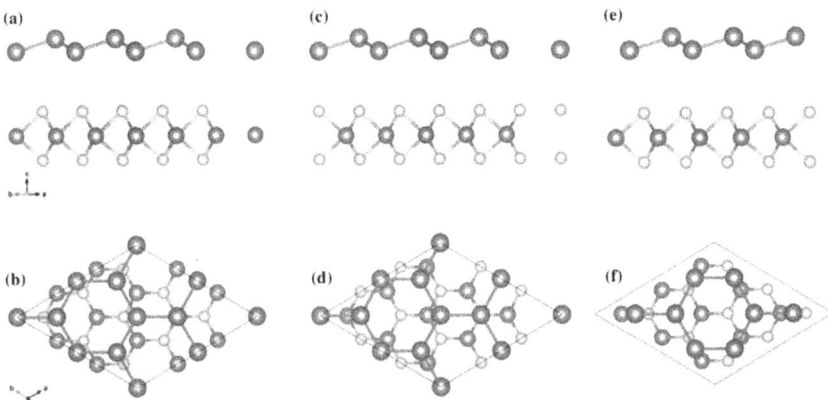

FIGURE 6.5   (a-f) Side views and top views of the three configurations of stanine/MoS$_2$ heterostructure. Reproduced with permission.[40] Copyright 2017, Springer US.

6.3.2 GROUP VA

6.3.2.1 Phosphorene-Based Heterostructures

As a member of the 2D material family, phosphoenes have been studied extensively because of their excellent electronic and optical properties. Many articles and comprehensive reviews have been published on the various phosphoene heterostructures. Phosphoene exhibits significant advantages such as thick-dependent band gaps, high carrier mobility, and high anisotropy. However, the instability in the air is the main obstacle in its application. Therefore, it is necessary to improve it through certain means. A good strategy is to protect phosphoene by covering it with other 2D materials. It was found that covering phosphoene with graphene, h-BN, and MoS_2 could improve the stability of the phosphoene without reducing its electrical properties.[41–44] In the phosphoene/graphene heterostructure, the electronic structures of both phosphoene and graphene are well preserved due to the weak interaction between them. In the phosphorene/h-BN heterostructure, h-BN could effectively improve the performance of phosphoene-based FETs. Zhang et al. found that the interlayer distance, binding energy, and charge transfer in phosphorene/graphene and phosphorene/h-BN are similar.[45] The heterostructure retains several attractive features of phosphorene, including direct band gap and linear dichroism. However, the large redistribution of electrostatic potential at the interface may affect the exciton behavior of phosphorene. The graphene and h-BN can not only act as a coating to protect phosphorene, but also act as an active layer to regulate the carrier dynamics and optical properties of phosphorene.

It is well known that phosphorene has the potential to be used in high performance electronic devices. The p-type phosphorene and n-type monolayer MoS_2 can form a type II arrangement at the interface, as shown in Figure 6.6.[46, 47] Moreover, the p-n diode can be used as a photodetector with a maximum response rate of 418 mW^{-1} under the illumination of a 633 nm laser with a power of 1 μW, which is nearly 100 times higher than the reported phosphorene phototransistor.[47] Due to its excellent photoelectronic properties, phosphoene has the potential to be used in broad band and ultrafast optoelectronic applications, especially at infrared wavelengths. Ye et al. found that phosphorene/MoS_2 photodetectors can cover the visible to mid-infrared spectrum.[48]

Guo et al. found that the phosphorene/TMDC heterostructure is a semiconductor and the vertical electric field can reduce its band gap.[49] Wang et al., by studying the graphene/phosphoene/graphene heterostructures, found that these structures have weak thermal conductivity and their thermal transport barriers can be regulated by the number of layers in the heterostructures.[50] Yu et al. found that phosphoene/GeSe heterostructures have type II band alignment and indirect band gap. Moreover, by applying external strain, the indirect band-gap to direct band-gap and insulator-metal transitions can be achieved. Besides this, spontaneous electron-hole charge separation was also found. So, phosphoenes/ GeSe heterostructures have great potential for the application in optoelectronic devices.[51]

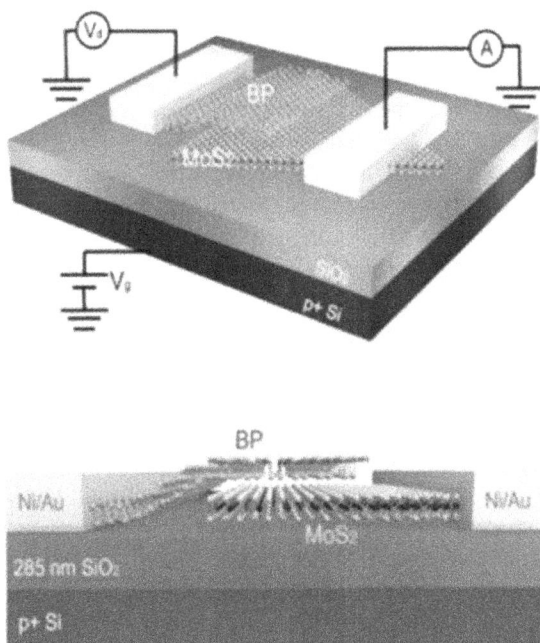

FIGURE 6.6 Schematic of phosphorene/MoS$_2$ heterostructure. Reproduced with permission.[46] Copyright 2014, American Chemical Society.

6.3.2.2 Arsenene-Based Heterostructures

Wang and Xia et al., while studying the structure and electronic properties of arsenene/graphene heterostructure, found that the linear Dirac-like dispersion relationship was still retained in the heterostructure.[52, 53] Moreover, by adjusting the interlayer distance, the transition from p-type Schottky barrier to n-type Schottky barrier can be realized. However, for arsenene/silicene heterostructure, the Schottky barrier is always p-type and the band gap is small, which can be opened by changing the interlayer distance. In the arsenene/FeCl$_2$ heterostructure, due to interfacial coupling, the spin splitting occurs at the bottom of the conduction band, and the maximum splitting energy is 123 meV.

Niu et al. revealed that arsenene/MoS$_2$, arsenene/tetracyano-quinodimethane, and arsenene/tetracyanonaphtho-quinodimethane heterostructures can form type II band alignments.[54] These arsenene-based heterostructures can not only satisfy all the requirements as photocatalysts for photocatalytic water splitting but can also show an excellent power conversion efficiency of ~20% as potential photovoltaics.

6.3.2.3 Antimonene-Based Heterostructures

Antimonene/silicon heterostructure exhibit a p-type Schottky barrier with a considerable band gap at the Dirac point.[52] The Schottky to Ohmic contact transition could be found by compressive strain applied to antimonene/graphene heterostructure. In particular, antimonene/germanene heterostructure has direct bandgap properties, which are completely different from the indirect bandgap characteristics of antimony and gapless characteristics of germanium.

Shao et al. investigated the biaxial strain effect on electronic structure tuning in antimonene-based van der Waals heterostructures through first-principles calculations.[55] By combining antimonene with graphene, arsenene, and h-BN, three new 2D van der Waals heterostructures, namely Sb/G, Sb/As, and Sb/h-BN, have been constructed, as shown in Figure 6.7. The results show that the band gap of Sb/h-BN and Sb/As heterostructures are 0.14 eV and 0.31eV, respectively, while the Sb/G heterostructures exhibit metallic properties. In addition, the strong interlayer coupling of Sb/h-BN and Sb/As can significantly regulate the band gap. Besides this, the biaxial strain can further adjust the electronic properties of 2D Sb/As and Sb/h-BN heterostructures, resulting in the gap transitions from indirect to direct. Interestingly, a continuous and controllable adjustment of the band gap from 1 eV to 0 eV can be achieved under biaxial strain. These results indicate that antimonene-based heterostructures have great potential in the application of infrared detectors and optoelectronic devices.

Zhou et al. systematically explored the electronic and topological properties of the antimonene/LaFeO$_3$ heterostructure through first-principles calculations.[56] The quantum spin–quantum anomalous Hall states can be observed in this heterostructure, and the band gap can be regulated by strain or electric field. Furthermore, they designed a device based on the quantum spin–quantum

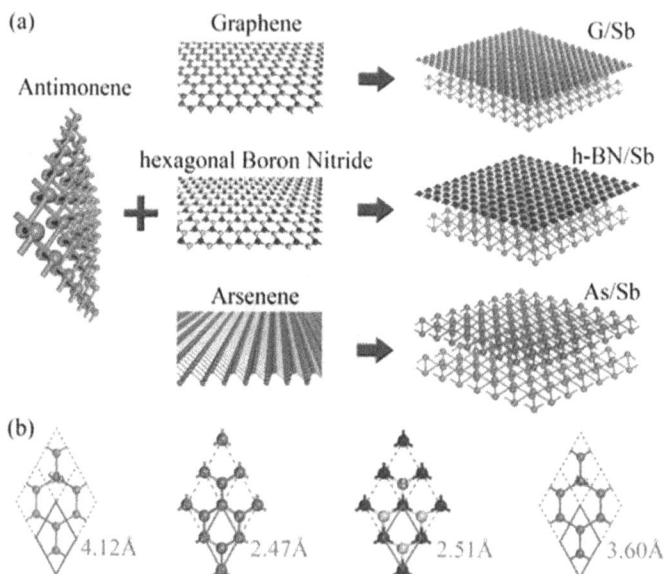

FIGURE 6.7 (a) Schematic diagrams of heterostructures based on antimonene. (b) The supercells used in calculations. Reproduced with permission.[55] Copyright 2016, Royal Society of Chemistry.

anomalous Hall heterostructure, which allows the carrier's degrees of freedom to be manipulated flexibly. These characteristics provide a promising way for its application in electronics, spintronics, and valley electronics.

6.3.2.4   Bismuthene-Based Heterostructure

Meysam et al. investigated the structural, electronic, mechanical, and optical properties of bismuthine/antimonene van der Waals heterostructure based on self-consistent density functional theory.[57] Several different Sb/Bi heterostructures have been explored and the most stable model has been found, as shown in Figure 6.8. Under the generalized gradient approximation, the most stable model is a semiconductor with an indirect band gap of 159 meV. However, when the spin orbit interaction is taken into account, the hesterostructure becomes a semimetallic. In addition, the results also show that the electronic properties of bismuthine/antimonene heterostructure are robust against the external electric field and biaxial strain. The heterostructure shows good mechanical properties with Young's modulus of 64.3 N/m. The optical properties of the bismuthine/antimonene heterostructures are very similar to those of monolayer antimonene, and are completely dependent on the polarization of incident light. The high structural stability, the electrons robustness to electric fields and strains, and the polarization-dependent optical properties of bismuthine/antimonene heterostructure give them great potential for use in splitters and nanoscale mirrors.

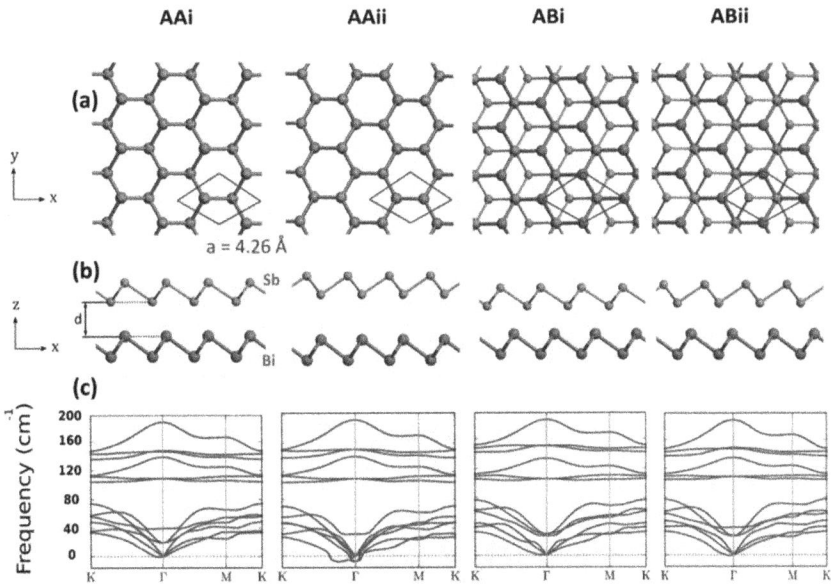

FIGURE 6.8   Four models of Bi/Sb herterostructure. (a) The top view, (b) the side view, and (c) the phonon dispersion for each model. Reproduced with permission.[57] Copyright 2020, Elsevier.

Sarker et al. proposed and investigated bismuthine/SiC heterostructure by first-principles calculations.[58] The results show that the electronic structure of heterostructure can be adjusted effectively by changing the interlayer distance and applying biaxial strain. Furthermore, with the increase of layer spacing, the influence of SiC on the bismuthine gradually decreases, so that the band gap will also decrease. The band gap is adjustable in the range of 0.48 eV to 0.59 eV.

Reis et al. synthesized bismuthine/SiC heterostructure by growing bismuthene on silicon carbide (0001) substrates.[59] In this heterostructure, the substrate not only stabilizes the quasi-2D topological insulation state, but also plays an important role in achieving large gaps. This approach provides a new approach for the study of large bandgap quantum spin Hall systems at room temperature.

6.3.3 GROUP IIIA

6.3.3.1 Boronene-Based Heterostructure

The borophene/graphene heterostructure exhibits ideal friction effects compared to almost all common low friction heterostructure in the study of Xu et al.[60] Jin et al. systematically investigated the transportation properties and electronic of borophene/phosphorene heterostructure through first-principles calculations. The calculated results show that the band gap can be tunneled through the interlayer spacing of heterostructures. When the layer spacing is 2.65Å, the borophene/phosphorene heterostructure electrode shows the best transportation performance.[61]

Hou et al. reported the borophene/graphene heterostructure humidity sensor exhibits long-time stability, fast response, and superhigh sensitivity; the structure is shown in Figure 6.9.[62] The sensitivity of the transducer is the highest among all the reported chemiresistive transducers based on 2D materials. At the relative humidity of 85% RH, the sensitivity of the manufactured borophene-based transducer is almost 700 times higher than that of graphene. Furthermore, the

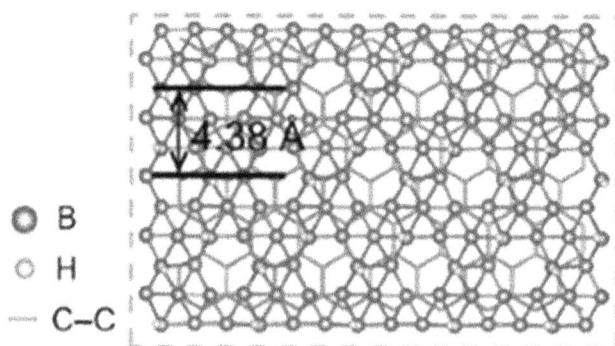

FIGURE 6.9 The structure of borophene-graphene heterostructure. Reproduced with permission.[62] Copyright 2020, Tsinghua University Press.

performance of the borophene/graphene flexible transducer keeps great stability after bending, which indicates the great potential of borophene-based heterostructures in the application of wearable electronics.

Bian et al. investigated the electronic and optical properties of borophene/MoS_2 heterostructure.[63] The results indicate that the electronic properties can be regulated by the stack strategy of 1/6 vacancy borophene β_{12} and MoS_2. Furthermore, compared with MoS_2, the stacked heterostructure expands the range of light response. The stack could change the electronic and optical properties of MoS_2. These results provide a solution for the design of optoelectronic devices with tunable Schottky contacts and good optical properties.

6.3.3.2 Aluminene-Based Heterostructure

Ravindra et al. revealed that the interaction between aluminene and the BN or graphene is weak and dominated by the van der Waals forces, through first-principles calculations; the structure is shown in Figure 6.10.[64] The calculated electronic structures show that the aluminene/graphene, aluminene/BN, and graphene/aluminene/graphene heterostructure are metallic. However, BN/aluminene/BN heterostructure is predicted to be a semiconductor. The results show that the interaction between Al atoms with N atoms promotes the opening of the gap in BN/aluminene/BN heterostructures. The properties of the electronic tunnel were investigated for the three-layer heterostructure and the stacking-dependence of the electronic structure was confirmed. The three-layer heterostructures exhibit a strong dependence between the stacking arrangements and the electronic structure. In addition, Dirac cones are observed for AB'A-stacked configuration of BN/aluminene/BN. For the AAA-, AB'A-, and ABC-configurations of BN/aluminene/BN, the band gaps are found to be 0.22, 0.96, and 0.07 eV, respectively.

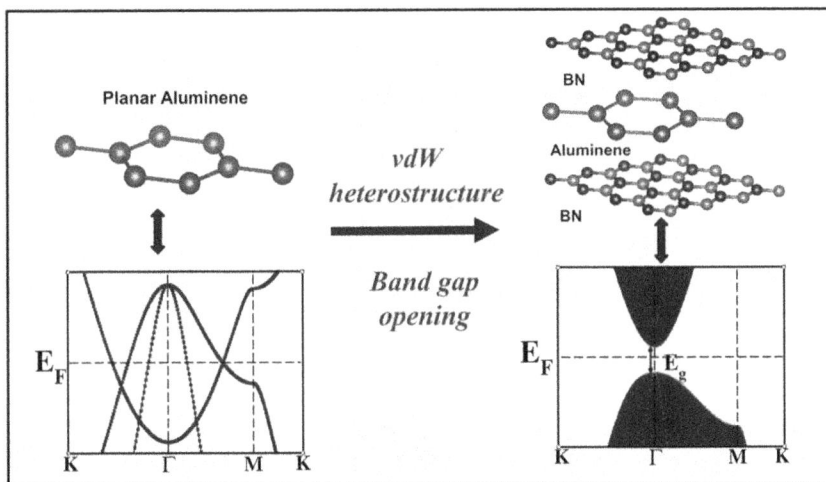

FIGURE 6.10 The structure of BN/Alumlnene/BN. Reproduced with permission.[64] Copyright 2020, Elsevier.

6.3.4   GROUP VIA

6.3.4.1   Tellurene-Based Heterostructure

Due to the unique structure of tellurene, it is difficult to form a lateral heterostructure. So, all the research about tellurene heterostructure is mainly regarding the van der Waals heterostructure.

Tellurene has been considered as an ideal candidate material for the design of high-efficiency heterostructure solar cells due to its excellent photoelectronic properties, high stability, and strong visible light absorption. Yang et al. found that tellurene/TMD heterostructure presented an ideal band gap of 1.47 eV.[65] Moreover, tellurene/TMD heterostructures exhibit desirable type II band alignment and excellent photoelectronic properties, such as strong visible light absorption of 5.0×10^5 cm^{-1} and high carrier mobility of 2.87×10^3 cm^{2}V^{-1}s^{-1}. Furthermore, the results show that the maximum power conversion efficiency of the proposed tellurene/WTe$_2$ and tellurene/MoTe$_2$ heterostructure solar cells can reach 22.5% and 20.1%, respectively.

Lin et al. investigated the properties of tellurene/VS$_2$ heterostructure through first-principles calculations; the structure is shown in Figure 6.11.[66] The results show that the tellurium/VS$_2$ heterostructure is ferromagnetic. The ferromagnetic properties of 2D tellurene/VS$_2$ heterostructures provide the possibility for the realization of high temperature ferromagnetic heterostructures. Furthermore, the exciton behavior of tellurium/VS$_2$ heterostructure may be quite different from that of the isolated tellurene, because the gradient of the interface potential may facilitate the separation of electron and hole.

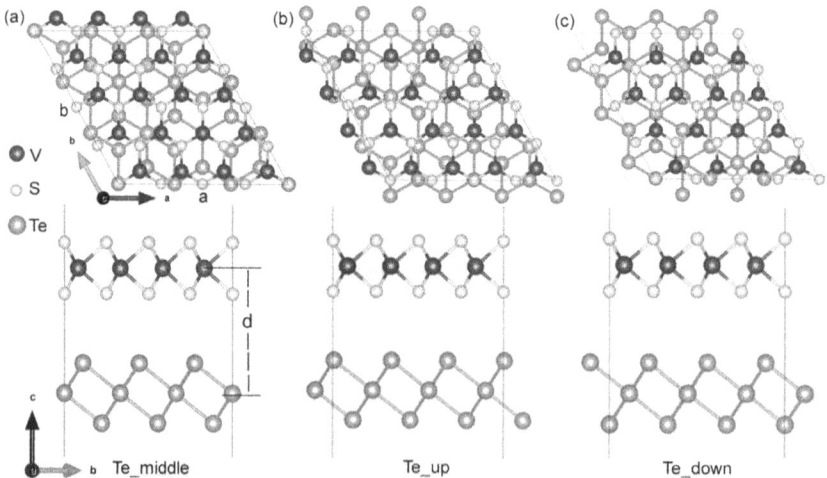

FIGURE 6.11   Three stacking structures of tellurene/VS$_2$. Reproduced with permission.[66] Copyright 2020, Elsevier.

6.4 LATERAL HETEROSTRUCTURE

In addition to the 2D van der Waals heterostructure that has been widely studied, at the same time, 2D lateral heterostructure has also attracted extensive attention of researchers. The lateral heterostructures are considered to be easier for planner integration and exhibit unique electronic and photoelectronic properties.

The 2D lateral heterostructure of arsenene/antimonene have been investigated by Sun et al.[67] The 2D lateral arsenene/antimonene heterostructure exhibits a direct energy gap without any regulation compared with the original serrated As and Sb monolayers. These lateral heterostructures connected by covalent bonds tend to form a type-II band alignment, which can promote carrier separation. In addition, further studies have shown that the arsenene/antimonene heterostructure has width-independent high carrier mobility. These properties make lateral heterostructure materials have great potential in the application of electronics and optoelectronics fields. Therefore, the fabrication of arsenene/antimonene lateral heterostructures not only enriches the family of new heterostructures, but also provides a new strategy for the manufacture of future equipment.

Accurate spatial control of the electrical properties of thin films is a key capability for modern integrated circuits. Graphene/h-BN has attracted a lot of attention due to its ability to allow different atomic composition to coexist in a continuous atomic film.[68, 69] With appropriate controls, the band gaps and magnetic properties can be engineered. Mark P et al. reported a versatile and scalable process that allows for the spatially controlled synthesis of graphene/h-BN lateral heterostructures.[70] Furthermore, the results of conductance measurements revealed the laterally insulating behavior of h-BN regions and the low sheet resistances and high carrier mobilities behavior of graphene regions.

2D heterostructures have attracted much attention due to their unique properties and potential applications in modern electronics and optoelectronics. Pan et al. reported a one-step CVD method for the growth of band alignment continuously modulated WS_2–$WS_{2(1-x)}Se_{2x}$ ($0 < x \leq 1$) monolayer lateral heterostructures,[71] as shown in Figure 6.12. Local photoluminescence (PL), Raman measurements, and Kelvin probe force microscopy (KPFM) investigations revealed the position-dependent composition, band gap information on the nanosheets, and the tunable band alignments in the heterostructures. The direct growth of high-quality atomic-level heterostructures with controllable band alignment is of great significance for the application of 2D semiconductors in integrated electronic and optoelectronic devices. Besides this, the strain or dielectric modulation has been used to construct van der Waals heterostructures, but there are very few works about lateral heterostructures published so far. So, there is still a lot of work that needs to be done.

6.5 CONCLUSION

This chapter reviewed some of the important literature on 2D Xene based heterostructures to emphasize the recent research in the structural characteristics

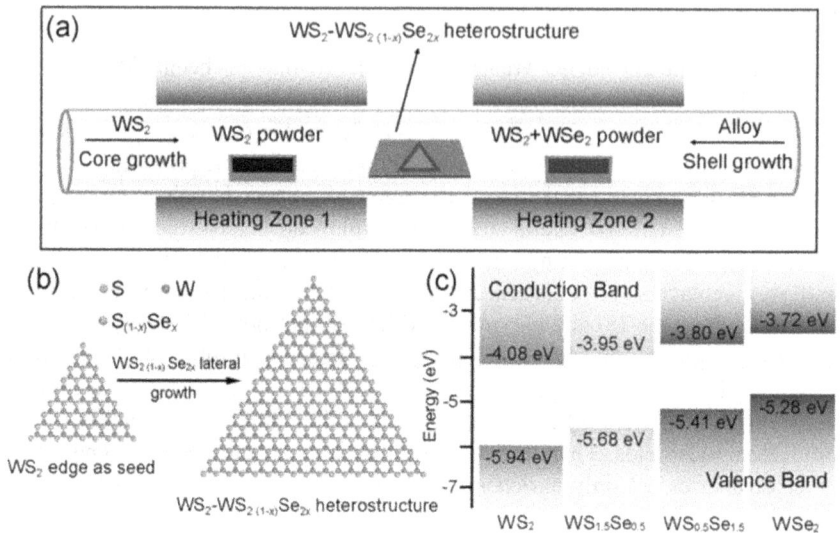

FIGURE 6.12 (a) Schematic of chemical vapor deposition setup. (b) Schematic illustration of the synthesis of the WS_2–$WS_{2(1-x)}Se_{2x}$ heterostructures. (c) Calculated band structures of the $WS_{2(1-x)}Se_{2x}$ with different Se mole fraction (x). Reproduced with permission.[71] Copyright 2018, American Chemical Society.

and basic properties of heterostructures and the applications in optoelectronics, energy storage field effect transistors, and photodetectors. The novel physical and mechanical properties of 2D monelemental materials hint at the great potential of it in 2D heterostructures. Heterostructures not only could overcome the inherent limitations of each material, but also could produce many interesting physicochemical properties through their appropriate combination.

All these studies will facilitate the design of van der Waals heterostructures and lateral heterostructures, and will be of critical significance for the applications of 2D monelemental heterostructure materials.

REFERENCES

1. Hu, L. et al., Phosphorene/ZnO nano-heterojunctions for broadband photonic non-volatile memory applications. *Advanced Materials*, 30, 30, 2018.
2. Long, G. et al., Achieving ultrahigh carrier mobility in two-dimensional hole gas of black phosphorus. *Nano Letters*, 16, 12, 2016.
3. Dean, C. R. et al., Boron nitride substrates for high-quality graphene electronics. *Nature Nanotechnology*, 5, 10, 2010.
4. Haigh, S. et al., Cross-sectional imaging of individual layers and buried interfaces of graphene-based heterostructures and superlattices. *Nature Materials*, 11, 9, 2012.
5. Wang, D. et al., Thermally induced graphene rotation on hexagonal boron nitride. *Physical Review Letters*, 116, 12, 2016.
6. Roy, K. et al., Graphene–MoS 2 hybrid structures for multifunctional photoresponsive memory devices. *Nature Nanotechnology*, 8, 11, 2013.

7. Choi, M. S. et al., Controlled charge trapping by molybdenum disulphide and graphene in ultrathin heterostructured memory devices. *Nature Communications*, 4, 1, 2013.
8. Lee, G.-H. et al., Flexible and transparent MoS2 field-effect transistors on hexagonal boron nitride-graphene heterostructures. *ACS Nano*, 7, 9, 2013.
9. Liu, Z. et al., Direct growth of graphene/hexagonal boron nitride stacked layers. *Nano Letters*, 11, 5, 2011.
10. Lin, Y.-C. et al., Direct synthesis of van der Waals solids. *Acs Nano*, 8, 4, 2014.
11. Oshima, C. et al., A hetero-epitaxial-double-atomic-layer system of monolayer graphene/monolayer h-BN on Ni (111). *Solid State Communications*, 116, 1, 2000.
12. Ding, X., Ding, G., Xie, X., Huang, F. & Jiang, M., Direct growth of few layer graphene on hexagonal boron nitride by chemical vapor deposition. *Carbon*, 49, 7, 2011.
13. Shi, Y. et al., van der Waals epitaxy of MoS2 layers using graphene as growth templates. *Nano Letters*, 12, 6, 2012.
14. Ago, H. et al., Controlled van der Waals epitaxy of monolayer MoS2 triangular domains on graphene. *ACS Applied Materials & Interfaces*, 7, 9, 2015.
15. Shi, J. et al., All chemical vapor deposition synthesis and intrinsic bandgap observation of MoS2/graphene heterostructures. *Advanced Materials*, 27, 44, 2015.
16. Rossi, A. et al., Patterned tungsten disulfide/graphene heterostructures for efficient multifunctional optoelectronic devices. *Nanoscale*, 10, 9, 2018.
17. Bianco, G. et al., Direct epitaxial CVD synthesis of tungsten disulfide on epitaxial and CVD graphene. *RSC Advances*, 5, 119, 2015.
18. Shim, G. W. et al., Large-area single-layer MoSe2 and its van der Waals heterostructures. *ACS Nano*, 8, 7, 2014.
19. Lin, Y.-C. et al., Atomically thin heterostructures based on single-layer tungsten diselenide and graphene. *Nano letters*, 14, 12, 2014.
20. Lin, Y.-C. et al., Tuning electronic transport in epitaxial graphene-based van der Waals heterostructures. *Nanoscale*, 8, 16, 2016.
21. Ugeda, M. M. et al., Characterization of collective ground states in single-layer NbSe 2. *Nature Physics*, 12, 1, 2016.
22. Ben Aziza, Z. et al., van der Waals epitaxy of GaSe/graphene heterostructure: electronic and interfacial properties. *ACS Nano*, 10, 10, 2016.
23. Li, X. et al., van der Waals epitaxial growth of two-dimensional single-crystalline GaSe domains on graphene. *ACS Nano*, 9, 8, 2015.
24. Dang, W., Peng, H., Li, H., Wang, P. & Liu, Z., Epitaxial heterostructures of ultrathin topological insulator nanoplate and graphene. *Nano Letters*, 10, 8, 2010.
25. Garcia, J. M. et al., Graphene growth on h-BN by molecular beam epitaxy. *Solid State Communications*, 152, 12, 2012.
26. Roth, S., Matsui, F., Greber, T. & Osterwalder, J., Chemical vapor deposition and characterization of aligned and incommensurate graphene/hexagonal boron nitride heterostack on Cu (111). *Nano Letters*, 13, 6, 2013.
27. Roth, S., Greber, T. & Osterwalder, J.r., Some like it flat: Decoupled h-BN monolayer substrates for aligned graphene growth. *ACS Nano*, 10, 12, 2016.
28. Solís-Fernández, P., Bissett, M. & Ago, H., Synthesis, structure and applications of graphene-based 2D heterostructures. *Chemical Society Reviews*, 46, 15, 2017.
29. Tan, H. et al., Doping graphene transistors using vertical stacked monolayer ws2 heterostructures grown by chemical vapor deposition. *ACS Applied Materials & Interfaces*, 8, 3, 2016.
30. Jiang, L. et al., Monolayer MoS2–graphene hybrid aerogels with controllable porosity for lithium-ion batteries with high reversible capacity. *ACS Applied Materials & Interfaces*, 8, 4, 2016.

31. Chang, K. & Chen, W., L-cysteine-assisted synthesis of layered MoS2/graphene composites with excellent electrochemical performances for lithium ion batteries. *Acs Nano*, 5, 6, 2011.

32. Wang, Z. et al., CTAB-assisted synthesis of single-layer MoS 2–graphene composites as anode materials of Li-ion batteries. *Journal of Materials Chemistry A*, 1, 6, 2013.

33. Li, H. et al., MoS2/graphene hybrid nanoflowers with enhanced electrochemical performances as anode for lithium-ion batteries. *The Journal of Physical Chemistry C*, 119, 14, 2015.

34. Guo, G.-C. et al., First-principles study of phosphorene and graphene heterostructure as anode materials for rechargeable Li batteries. *The Journal of Physical Chemistry Letters*, 6, 24, 2015.

35. Li, G. et al., Stable Silicene in Graphene/Silicene van der Waals Heterostructures. *Advanced Materials*, 30, 49, 2018.

36. Shi, L., Zhao, T. S., Xu, A. & Xu, J. B., Ab initio prediction of a silicene and graphene heterostructure as an anode material for Li- and Na-ion batteries. *Journal of Materials Chemistry A*, 4, 42, 2016.

37. Chen, X. et al., The electronic and optical properties of novel germanene and antimonene heterostructures. *Journal of Materials Chemistry C*, 4, 23, 2016.

38. Zhao, N. & Schwingenschlögl, U., Transition from Schottky to Ohmic contacts in Janus MoSSe/germanene heterostructures. *Nanoscale*, 12, 21, 2020.

39. Noshin, M., Khan, A. I. & Subrina, S., Thermal transport characterization of stanene/silicene heterobilayer and stanene bilayer nanostructures. *Nanotechnology*, 29, 18, 2018.

40. Liang, D. et al., Tunable band gaps in stanene/MoS 2 heterostructures. *Journal of Materials Science*, 52, 10, 2017.

41. Sinha, S., Takabayashi, Y., Shinohara, H. & Kitaura, R., Simple fabrication of air-stable black phosphorus heterostructures with large-area hBN sheets grown by chemical vapor deposition method. *2D Materials*, 3, 3, 2016.

42. Kuriakose, S. et al., Black phosphorus: Ambient degradation and strategies for protection. *2D Materials*, 5, 3, 2018.

43. Gamage, S. et al., Reliable passivation of black phosphorus by thin hybrid coating. *Nanotechnology*, 28, 26, 2017.

44. Constantinescu, G. C. & Hine, N. D., Multipurpose black-phosphorus/hBN heterostructures. *Nano letters*, 16, 4, 2016.

45. Cai, Y., Zhang, G. & Zhang, Y.-W., Electronic properties of phosphorene/graphene and phosphorene/hexagonal boron nitride heterostructures. *The Journal of Physical Chemistry C*, 119, 24, 2015.

46. Deng, Y. et al., Black phosphorus–monolayer MoS2 van der Waals heterojunction p–n diode. *ACS Nano*, 8, 8, 2014.

47. Huang, L. et al., Electric-field tunable band offsets in black phosphorus and MoS2 van der Waals pn heterostructure. *The Journal of Physical Chemistry Letters*, 6, 13, 2015.

48. Ye, L., Li, H., Chen, Z. & Xu, J., Near-infrared photodetector based on MoS2/black phosphorus heterojunction. *Acs Photonics*, 3, 4, 2016.

49. Guo, H., Lu, N., Dai, J., Wu, X. & Zeng, X.C., Phosphorene nanoribbons, phosphorus nanotubes, and van der Waals multilayers. *The Journal of Physical Chemistry C*, 118, 25, 2014.

50. Wang, X., Zebarjadi, M. & Esfarjani, K., First principles calculations of solid-state thermionic transport in layered van der Waals heterostructures. *Nanoscale*, 8, 31, 2016.

51. Yu, W. et al., Tunable electronic properties of GeSe/phosphorene heterostructure from first-principles study. *Applied Physics Letters*, 109, 10, 2016.
52. Wang, Y. & Ding, Y., The electronic structures of group-V–group-IV hetero-bilayer structures: A first-principles study. *Physical Chemistry Chemical Physics*, 17, 41, 2015.
53. Xia, C., Xue, B., Wang, T., Peng, Y. & Jia, Y., Interlayer coupling effects on Schottky barrier in the arsenene-graphene van der Waals heterostructures. *Applied Physics Letters*, 107, 19, 2015.
54. Niu, X., Li, Y., Zhou, Q., Shu, H. & Wang, J., Arsenene-based heterostructures: highly efficient bifunctional materials for photovoltaics and photocatalytics. *ACS Applied Materials & Interfaces*, 9, 49, 2017.
55. Lu, H., Gao, J., Hu, Z. & Shao, X., Biaxial strain effect on electronic structure tuning in antimonene-based van der Waals heterostructures. *RSC Advances*, 6, 104, 2016.
56. Zhou, T. et al., Quantum spin–quantum anomalous Hall effect with tunable edge states in Sb monolayer-based heterostructures. *Physical Review B*, 94, 23, 2016.
57. Mozvashi, S. M., Vishkayi, S. I. & Tagani, M. B., Antimonene/bismuthene vertical van-der waals heterostructure: a computational study. *Physica E: Low-dimensional Systems and Nanostructures*, 118, 2020.
58. Sarker, J.D. et al., Tunable electronic properties in bismuthene/2D silicon carbide van der Waals heterobilayer. *Japanese Journal of Applied Physics*, 59, SC, 2019.
59. Reis, F. et al., Bismuthene on a SiC substrate: A candidate for a high-temperature quantum spin Hall material. *Science*, 357, 6348, 2017.
60. Xu, Y., Cheng, Z., Zhu, X., Lu, Z. & Zhang, G., Ultra-low friction of Graphene/ Honeycomb Borophene Heterojunction. *Tribology Letters*, 69, 2, 2021.
61. Jin, J. et al., Study on the transport properties of borophene/phosphorene hetero-junctions. *Emerging Materials Research*, 9, 3, 2020.
62. Hou, C., Liu, B., Wu, Z. & Yin, Y., Borophene-graphene heterostructure: Preparation and ultrasensitive humidity sensing. *Nano Research*, 14, 2020.
63. Wei, J., Li, W., Liao, B. & Bian, B., Electronic and optical properties of vertical borophene/MoS2 heterojunctions. *Materials Chemistry and Physics*, 252, 2020.
64. Pandey, D., Kumar, A., Chakrabarti, A. & Pandey, R., Stacking-dependent electronic properties of aluminene based multilayer van der Waals heterostructures. *Computational Materials Science*, 185, 2020.
65. Wu, K., Ma, H., Gao, Y., Hu, W. & Yang, J., Highly-efficient heterojunction solar cells based on two-dimensional tellurene and transition metal dichalcogenides. *Journal of Materials Chemistry A*, 7, 13, 2019.
66. Yu, W. et al., Ferromagnetic half-metal properties of two dimensional vertical tel-lurene/VS2 heterostructure: A first-principles study. *Computational Materials Science*, 171, 2020.
67. Sun, Q. et al., Design of lateral heterostructure from arsenene and antimonene. *2D Materials*, 3, 3, 2016.
68. Zhao, L. et al., Visualizing individual nitrogen dopants in monolayer graphene. *Science*, 333, 6045, 2011.
69. Ci, L. et al., Atomic layers of hybridized boron nitride and graphene domains. *Nature Materials*, 9, 5, 2010.
70. Levendorf, M. P. et al., Graphene and boron nitride lateral heterostructures for atomically thin circuitry. *Nature*, 488, 7413, 2012.
71. Zheng, B. et al., Band alignment engineering in two-dimensional lateral hetero-structures. *Journal of the American Chemical Society*, 140, 36, 2018.

7 Defective 2D Xenes materials

Huating Liu, Zongyu Huang, and Jianxin Zhong

CONTENTS

7.1 INTRODUCTION

The successful exfoliation of graphene, the first material found to consist of a single layer of atoms, from graphite has opened the door to a two-dimensional (2D) material world.[1] Graphene has many excellent characteristics, such as good electrical conductivity,[2] good thermal conductivity,[3] good lubrication performance,[4] high tensile strength,[5] strong sensing ability,[6] and so on.[7] It has become a hot and emerging field in the field of physics and materials science.[8] However, with the continuous progress of research on graphene, it has been found that graphene has the risk of environmental pollution, and it is very difficult and costly to produce graphene on a large scale.[9, 10] More importantly, the zero-band gap feature of graphene has greatly limited its application in semiconductor materials.[11, 12] In order to open the band gap in graphene, researchers have made many efforts.[13–15]

In the process of material preparation, imperfect preparation technology will inevitably lead to structural defects or bring doped atoms into the impurities.[16–18] Kislenko et al. carried out a theoretical study on the influence of defects on electron transfer at the interface of graphene solution, and the calculation results showed that defects could increase the charge transfer rate in graphene by an order of magnitude, and the change of defect type could selectively promote

DOI: 10.1201/9781003207122-7

electron transfer to a certain reagent.[19] Chen et al. pointed out that the appearance of graphene defects changed the length of the valence bonds between atoms, and also changed the type of hybrid orbitals of some carbon atoms, resulting in changes in the electronic properties of graphene defect regions.[20] The influence of external atom introduction defects on the electronic properties of graphene is more complex and interesting.[21, 22] The internal and external atomic substitution defects formed by nitrogen and boron atoms cause resonance scattering effect on graphene, which further affects the electronic properties of graphene.[23] Defective graphene provides the possibility to realize selective electrochemical catalysis.[24] By effectively controlling the distribution of defects, a variety of 2D crystal structures based on graphene could be obtained, providing more new ideas for the performance regulation of graphene.[25–27]

Even with so many advantages, the defect graphene, as a 2D monoelement material, should have rich physical properties from borophene to tellurene, among which black phosphorene (BP) as the representative has become a research hotspot in recent years.[28–30] The BP has a band gap value adjustable with the number of layers from 0.3 eV in the bulk to 2.0 eV in monolayer, allowing it to absorb wavelengths in the visible range and the infrared range for communication, and its carrier mobility is also high.[31–33] Therefore, BP has broad application prospects in the fields of transistors, energy storage, supercapacitors, memory components, and solar power generation.[34, 35] More importantly, just as defects bring rich properties to graphene, defective phosphoenes can not only effectively improve its environmental instability, enhance activation sites, enhance catalytic performance, but also achieve conversion from metal, semiconductor, semi-metal, etc., when doped with different external atoms.[36, 37] Meanwhile, defective phosphoenes will enhance the absorption, detection, and storage capacity of various gas molecules.[38, 39]

Therefore, through this chapter, it is necessary to make a simple and as comprehensive as possible summary of the properties of monoelemental materials under defect engineering studied so far, including monoelemental materials from group IIIA ~VIA. We will summarize the classification of defects, the properties of monoelemental materials with different groups and different defect types, and their potential applications. From this, we can find out the shortcomings of the research and further improve and strengthen them. Meanwhile, we can inspire research ideas and improve experimental methods, so as to provide more space and ideas for the practical application of single-element materials.

7.2 CLASSIFICATION AND INTRODUCTION OF XENES DEFECTS

It is well known that the atoms in an ideal crystal are arranged in regular, periodic order in space. However, due to the influence of the thermal movement of atoms and other conditions, the arrangement of some atoms in the crystal lattice will deviate from the ideal situation, thus destroying the symmetry of the crystal and causing some defects. As a member of the family of 2D materials, Xenes materials also have abundant types of defects. According to the formation of defects, they can be divided into intrinsic defects and impurity defects. At the same time, the methods to realize defects are introduced.

7.2.1 INTRINSIC DEFECTS

Intrinsic defect, as the name implies, is a defect occurring in the structure of the material itself. According to the change of hybridization form of the orbital between atoms, some atoms will be missing or extra in the system. For 2D Xenes materials, we mainly consider single-vacancy defects, multi-vacancy defects, Stone-Wales (SW) defects, line defects, and so on. As shown in Figure 7.1(a),

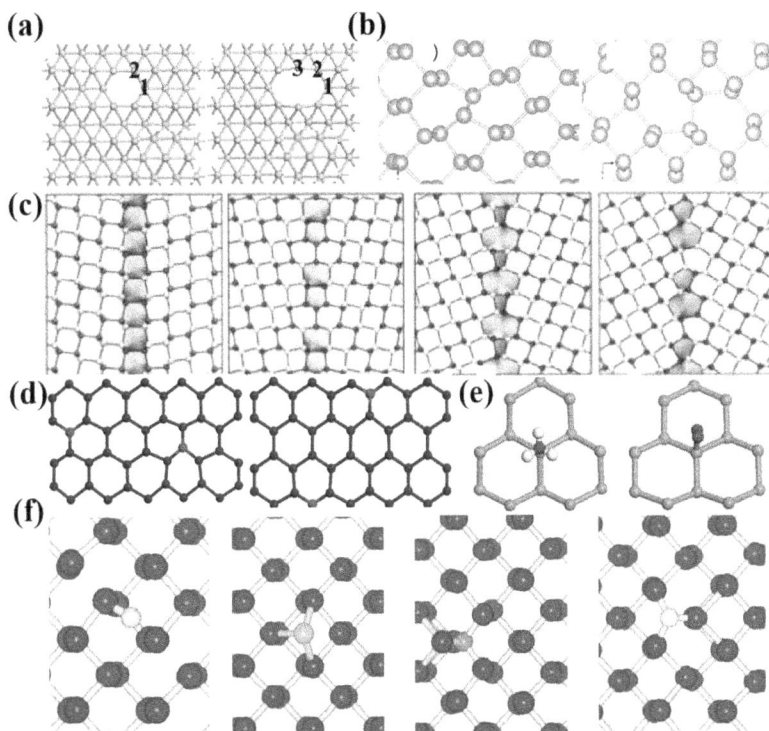

FIGURE 7.1 (a) The atomic structure (upper panel) for several edge atoms of monovacancies (MV) and divacancies (DV) borophene sheets. Reproduced with permission.[40] Copyright 2018, Royal Society of Chemistry. (b) Top view of Stone-Wales–like defects in phosphorene. Reproduced with permission.[41] Copyright 2018, Elsevier. (c) Structure of grain boundaries at various misorientation angles of phosphorene sheets. Reproduced with permission.[42] Copyright 2014, American Chemical Society. (d) Schematic representation of doped stanene with single doping and edge doping patterns. Reproduced with permission.[45] Copyright 2018, Royal Society of Chemistry. (e) Atomic structures of the adsorption of gas molecules in the most stable configuration on germanene (NH_3 and NO). Reproduced with permission.[43] Copyright 2014, Royal Society of Chemistry. (f) Top views of equilibrium (optimized) atomic structure of the monolayer antimonene phase after the adsorption of adatoms. Reproduced with permission.[44] Copyright 2016, American Physical Society.

when an atom is lost in the continuous arrangement of atomic structures and vacancies appear, the atoms around the system will get closer or disperse to form new covalent bonds and lattice distortion will occur. Losing more atoms in addition to a single vacancy will result in a multi-vacancy defect (Figure 7.1(a)).[40] Stone-Wales defects are formed when two atoms rotate and change the connecting bonds between the atoms without atom introduction or removal, as shown in Figure 7.1(b).[41] The single element materials with different crystal orientations have line defects at the intersection of edges due to the different initial crystal orientations (Figure 7.1(c)).[42]

7.2.2 Impurity Defects

Impurity defects are the defects caused by the introduction of external impurities. Generally, their capacity and size are small. According to the different positions of doped elements, they can be divided into out-of-plane adsorption and in-plane substitution. Due to the different properties of impurity atoms and intrinsic atoms, the regular arrangement of atoms will be broken and the periodic potential field around them will change.

Out-of-plane heteroatom adsorption defects: Under the condition of chemical vapor deposition or strong oxidation, due to the use of metal elements or oxygen-containing oxidants in the process, the surface of 2D materials will inevitably introduce metal atoms or oxygen-containing functional groups. These heteroatoms bond with the atoms in the system by strong chemical bonds or weak vdWs forces, which constitute the heteroatomic introduction defects. In addition to metal atoms, it also includes oxygen atoms or oxygen-containing functional groups such as hydroxyl groups and carboxyl groups. According to the number of layers in these Xenes materials, the foreign atoms can be effectively adsorbed outside the layers or between the layers. In addition, according to the different crystal structure, the outer doped atom also has different adsorption sites such as top, vacancy, and bridge. The corresponding structural schematic diagram is shown in Figures 7.1 (e) and (f).[43, 44]

In-plane impurity atomic substitution defects: The positions of doped atoms in 2D materials can be replaced by atomic groups from the same or different groups. These impurity atoms constitute the in-plane heteroatomic substitution defect. In addition, by effective control of the method, not only single atom doping defects can exist in the system, but also multiple atoms can exist at the same time. For the substitution doping of the inner atoms, the site of substitution doping will be fully considered according to the problem of structural symmetry. As far as possible, a wealth of configurations available for substitution doping defects are given in Figure 7.1 (d).[45]

7.2.3 INTRODUCTION OF DEFECTS

It is easy to introduce various kinds of defects in the actual process of preparing 2D monoelement materials. Intrinsic defects can be introduced in single-element materials by plasma etching, ball milling, high temperature annealing, and ion beam techniques.[46–50] For plasma etching, the defect density and type can be adjusted effectively by changing the plasma type, intensity, position, and irradiation time. For the ball milling process, surface defects can be controlled by the difference of grinding power and time, but the types of internal defects are difficult to identify and effectively control. It is not easy to introduce any other impurities into the nanomaterials. With ion beam technology as a pure physical surface modulation technology, high controllability and precision can be achieved, so it has been widely used in the field of microelectronics.

The experimental methods for introducing impurity defects into monoelemental materials include gas phase conversion method, wet chemical functionalization approach, ion exchange method, and chemical etching, etc.[51–54] Among them, the chemical gas transport method can not only control the amount of doping, but also high crystallinity and doping uniformity, so gas phase conversion is the most widely used way. As a traditional method to introduce foreign defects, the insertion method achieves controllable structure of the material by inserting other atoms with smaller atomic radius into the gap of the 2D monoelemental material system and effectively improves the performance of the material. As a traditional method to introduce foreign defects, the intercalation achieves controllable structure of the material by inserting other atoms with smaller atomic radius into the gap of 2D monoelemental material system and effectively improves the performance of the material. In recent years, there are more and more new synthetic methods that introduce defects, such as the electrolytic method, the technique of liquid ammonia intercalation and pulsed-plasma method, etc. There are a variety of ways to introduce defects, as summarized in Table 7.1. However, the current

TABLE 7.1
Defect Types and the Corresponding Routine Experimental Preparation Methods

Defect Type	Experimental Methods
Intrinsic defects	Plasma etching
	Ball milling
	High temperature annealing
	Ion beam techniques
Impurity defects	Gas phase conversion method
	Wet chemical functionalization approach
	Ion exchange method
	Chemical etching

experimental preparation method is difficult to distinguish the types of defects in detail and control the number of defects well, so further exploration is needed.

7.3   OPTOELECTRONIC PROPERTIES AND APPLICATIONS OF DEFECTIVE XENES

Defects can effectively affect the physical properties of 2D materials, including heat conduction, resistance, optics, mechanical properties, and other indicators, and also affect the chemical properties of the material surface, such as battery performance, surface catalytic performance, etc. In this section, we will introduce the optoelectronic properties and related applications of defective single element materials with in the different groups.

7.3.1   GROUP IIIA OF DEFECTIVE 2D XENES MATERIALS

Up to now, the groups-IIIA of monoelementals materials including borophene, aluminene, gallenene, and indiene have been studied to a certain extent, but the research on their related defects has only been developed to borophene related defects, so this part takes defective borophene as a typical example to make a certain summary.

As reported by Kistanov et al., the metallicity of borophene can be effectively enhanced under vacancy defects,[55] and Jiang's research showed that the defective borophene suppressed shuttle effect and enhanced capacity retention.[56] Surprisingly, Zhang et al. pointed out that when the vacancy defect in the borophene becomes multivacancies, the defective borophene turns from metal to a semiconductor.[57] According to Arabha et al., the porous borophene can effectively control its thermo-mechanical properties, transforming the anisotropic borophene into a square quasi-isotropic structure.[58] The borophene with line defects can effectively adsorb nitrogen-containing gas molecules. Even at relatively low biases, the adsorption of these nitrogen-containing gas molecules can induce a significant change in the current of the defective borophene, and the magnitude of the current is significantly related to the type of adsorbed molecules.[59]

When borophene is doped with light elements such as beryllium, carbon, and fluorine atoms, an effective transition between semiconductor and metal will occur as well as absorption of light within the different parts of the electromagnetic spectrum, as shown in Figure 7.2(a).[60] Among them, the theoretical capacity and Li diffusion property of borophene modified by F atom can be improved effectively.[61] In Figure 7.2(c), the N-modified borophene is similar to the borophene after the adsorption of alkali metals, which can enhance the adsorption capacity of H_2.[62] Li-doped modified borophene also enhanced the adsorption of different gas molecules.[63] Lherbier et al. discovered that fully hydrogenated borophene, or borophane, would induce graphene-like Dirac cone structures with excellent electrical transport properties and high transparency.[62,64] However, when borophene is adsorbed by 3D TM atoms, ferromagnetic can be induced effectively,

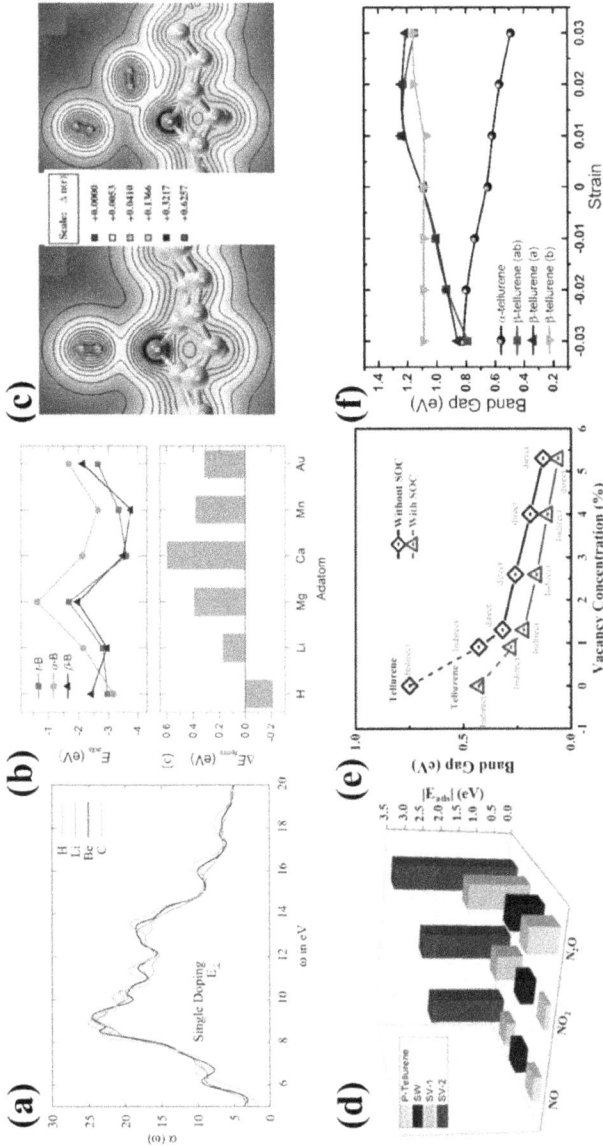

FIGURE 7.2 (a) Frequency-dependent absorption coefficient for perpendicular polarization single doping borophene. Reproduced with permission.[60] Copyright 2019, *Springer Nature*. (b) Substitutional doping in borophene: substitutional energies and the Fermi level shift induced by doping. Reproduced with permission.[66] Copyright 2017, Royal Society of Chemistry. (c) Charge density plot of 1H₂ molecule and 2H₂ molecules adsorbed over N-decorated borophene sheet. Reproduced with permission.[62] Copyright 2020, Elsevier. (d) Absolute values of E_{ads} of N_2O, NO and NO_2 adsorbed on the P-tellurene, SW defects, single-vacancy defects, respectively. Reproduced with permission.[135] Copyright 2021, Elsevier. (e) Band gap as a function of vacancy concentration intellurene. Reproduced with permission.[133] Copyright 2020, American Chemical Society. (f) Band gap as functions of the strain applied to tellurene sheets. Reproduced with permission.[134] Copyright 2018, Royal Society of Chemistry.

and the adsorption strength and Fermi energy level position can be changed effectively (Figure 7.2(b)).[65, 66] In addition, gas molecular functionalization of borophene can increase the electrical conductivity.

Based on the excellent properties of defective borophene mentioned above, it is possible to predict its many optoelectronic as well as energy storage and conversion applications, including the following: borophene with vacancy defects can be used as a field effect transistors (FETs), and borophene with line defects can be designed as a gas molecule sensor. Alkali-metal-modified borophene can effectively improve the storage capacity of hydrogen storage energy, and light-atom-doped borophene will improve the storage capacity and can be used as a candidate material for lithium or sulfur batteries.

7.3.2 GROUP IVA OF DEFECTIVE 2D XENES MATERIALS

In general, the research on the defects of silicene shows that the single-vacancy and double-vacancy defects can open the band gap of silicene (Figure 7.3(b)),[67] improve the mobility of the system, and promote the absorption of Na atoms and some gas molecules.[68] The defects of multi-vacancy clusters will lead to the decrease of the current of silicene.[69] Also, Kamyabmehr et al. pointed out that the existence of SW defects reduces the amount of reflectivity peaks, and the reflectivity at the plasma frequency drops sharply as shown in Figure 7.3(a).[70, 71] Hu et al. found that compared with graphene, porous silicene has a high selectivity to hydrogen of various gas molecules.[72] The corresponding energy spectrum changes with the adsorption height are shown in Figure 7.3(f).

When silicene is doped with nonmetallic atom P or N, metal semiconductor conversion occurs.[73] Chandiramouli et al. and Pablo-Pedro et al. respectively showed that when doped by Ge element of the same group, the adsorption characteristics of CO would be enhanced,[74] while when doped by C element, the Mott-Anderson conversion would be realized according to the different doping sites.[75] Besides this, when silicene is doped with metal atoms such as Pt, it has a strong catalytic potential for CO_2[76] (see Figure 7.3(c)), while when doped with Al atom, it can effectively regulate the band gap and dielectric function with various doping concentrations and strengthen the competitive advantage of the system over other one-atom-thick materials.[77] The impurity defects outside the silicene surface mainly revolve around passivation, including hydrogenated, fluorinated, oxidized, and brominated. The fully passivated silicene presents non-magnetic semiconductor characteristics, while the semi-passivated silicene presents ferromagnetic semiconductor or semi-metal behaviors (Figure 7.3(e)).[78] Different from passivation, Yao et al. reported that the band gap of Al atoms adsorbed silicene system increased. However, with the increase of shear deformation, the band gap of the adsorption system decreases gradually. At the same time, the shear deformation results in the red shift of the highest absorption peak and the highest reflection peak, and the degree of red shift increases with the increase of displacement.[79] Based on its effective current regulation, enhanced gas adsorption (Figure 7.3(d)) and controllable energy band change, defective silicone (including intrinsic vacancy defects and various element substitution dopants) can be used as a potential candidate material for gas separation and filtering applications, gas sensors, gas molecular catalysis, nonlinear light detectors, and other application fields.[80]

For germanene, single-vacancy defects maintain their Dirac cone features, while double-vacancy and SW defects open the band gap of germanene, and the band gap value is significantly affected by SOC effect.[81, 82] Meanwhile, the introduction of different vacancy defects will improve the quantum capacity of germanene.[83] Nonmetallic atom-doped germanenes such as B, N, and P are converted from semiconductors to metals and induced magnetic momentum.[84] For example, Ag, Au, Al, Cu, and Ti metal atom-doped germanene will be transformed into a quasi-metal. Germanene adsorption gas molecules (N_2, CO, CO_2, H_2O, NH_3, NO, NO_2, and O_2) can open the gap, while the high mobility of the Dirac cone is kept by the small organic molecules and alkali metals (Li, Na, and K).[85] In contrast, germanene decorated with ethynyl-derivatives (GeC_2X; X H, F, Cl, Br, I) will become a topological insulator with large bandgap.[86] When germanene adsorbed 3D TM atoms, it showed rich physical properties including non-magnetic metals, non-magnetic semiconductors, ferromagnetic metals, and ferromagnetic semiconductors according to different adsorbed atoms.[87] Therefore, defective germanene can be used as a very promising hydrogen and helium separation device, field effect transistor, etc.

Shen et al. reported that single-vacancy and SW defective stanene can open a direct band gap, and the size of the band gap is significantly affected by the SOC effect.[82] Moreover, the adsorption energy of stanene with intrinsic defects on gas molecules such as SO_2 and SO_3 is enhanced, which can effectively improve its performance as a gas sensor.[88] When the stanene is doped by the Al atom of group IIIA, the Fermi energy level will move to the valence band edge. When the stanene is doped by the fifth N and P atom of group VA, the Fermi energy level will move to the valence band edge.[89] When the stanene is co-doped by both Al and P atoms or Ga and As atoms, the system will open the band gap and turn into a semiconductor.[90, 91] Wang et al. showed that monolayer stanene doped with 4D transition metals (Mo, Nb, Rh, and Ru) exhibits metal or semiconductor properties depending on the doped element.[92] It is worth noting that Mo-doped stanene has high stability and good electrical conductivity, and shows excellent nitrogen reduction reaction (NRR) and can be used as a new targeted material for catalytic nitrogen reduction reaction.[93] Stanene doped by alkaline earth metal Be and Mg atoms will open the band gap when the doping concentration is 12.5%, and expand its feasibility as switch on/off devices.[94]

Due to the sizeable spin–orbit coupling, plumbene has attracted research interest. At present, the related defects found are the external defects introduced into plumbene by Hashemi et al. It is predicted that plumbene doped by 3D transition metal atoms can induce magnetic properties to achieve a theory room-temperature Curie temperature, with potential for magnetic storage and spintronic applications.[95] The research and exploration on the excellent properties of defective plumbene need to be further deepened and strengthened.

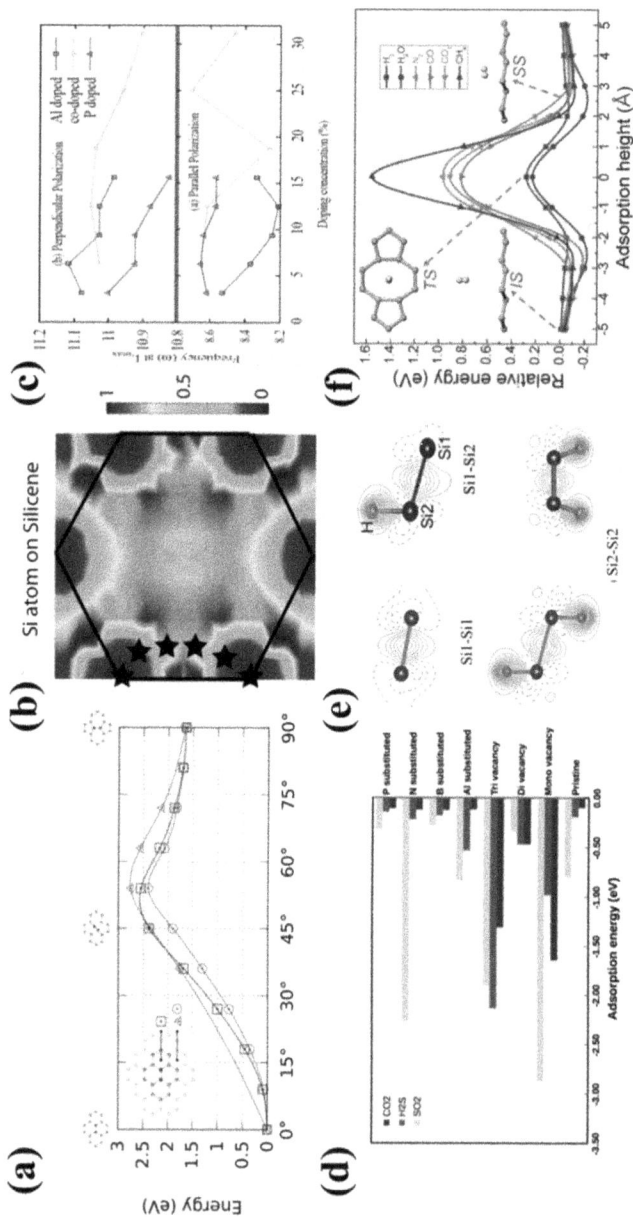

FIGURE 7.3 (a) Energetics of formation of monovacancy defect in silicene via the bond rotation. The black curve is obtained by relaxing the first-nearest hexagons to the silicene dimer in freestanding silicene layer. Red and blue curves are obtained by relaxing up to the second-nearest hexagons to the silicon dimer in the freestanding and supported silicene, respectively. Reproduced with permission.[71] Copyright 2013, American Physical Society. (b) The Si adatom on the silicene layer. The energy of the most favorable site is set to zero. Reproduced with permission.[67] Copyright 2013, American Physical Society. (c) Variation of static value of real part of the dielectric function with various doping concentrations for silicene. Reproduced with permission.[77] Copyright 2014, Royal Society of Chemistry. (d) van der Waals-induced adsorption energies

7.3.3 GROUP VA OF DEFECTIVE 2D XENES MATERIALS

Phosphorene is the most well-studied 2D monoelement material beyond graphene, and it is also the fastest to get practical application. Among them, the BP with saddle-like structures have received the most attention. Therefore, we mainly summarize the rich physical properties of BP, which takes its wide application in field effect transistors, photodetectors, sensors, batteries, hydrogen storage, and catalysis as examples to provide reference and guidance for the application of other Xenes materials.

The properties of BP under the intrinsic defect are rich: the band gap of single vacancy BP disappears,[96, 97] the band gap of double vacancy BP changes from direct to indirect, and the structure of SW BP changes (Figure 7.4(b)),[41] but it remains a semiconductor and the band gap increases. In general, for BP with the above-mentioned intrinsic defects, the carrier mobility is generally reduced, but the current increases and shows anisotropy, while the optical anisotropy is inhibited and the light absorption is quenched.[98] More importantly, BP with these intrinsic defects can effectively improve its performance as a battery anode material, as shown in Figure 7.4(a) and 7.4(c).[99, 100] The inherent defect engineered BP has high detection sensitivity and excellent reusable performance for many gas molecules like in Figure 7.4(d)[101] and organic molecules, including carbon dioxide (CO_2) molecules,[102] phosgene ($CoCl_2$) molecules,[103] formaldehyde (CH_2O) molecules,[104] cyanogen fluoride and chloride (CNCl) molecules.[105] Moreover, Pei et al. elucidated that different kinds of point defects change the reduction

---

FIGURE 7.3 (CONTINUED)

of incident gases on pure, defected, and substituted silicene. Reproduced with permission.[80] Copyright 2016, Royal Society of Chemistry. (e) Deformation charge density for three typical Si–Si bonds: Si1–Si1 bond in a silicene, Si1–Si2 bond in a half-hydrogenated silicene, and Si2–Si2 bonds in a full-hydrogenated silicene. The unhydrogenated/hydrogenated Si atoms are denoted as Si1 and Si2, respectively. Reproduced with permission.[78] Copyright 2016, Elsevier. (f) Energy profiles for H_2, H_2O, N_2, CO, CO_2, and CH_4 passing through the divacancy of silicene as a function of adsorption height. Reproduced with permission.[72] Copyright 2013, Royal Society of Chemistry. Figure 7.4 (a) Charge density distribution for point-defect phosphorene. Reproduced with permission.[99] Copyright 2020, Elsevier. (b) Bond length distribution for pristine, SW defected phosphorene, and N-doped SW defected phosphorene. Reproduced with permission.[41] Copyright 2018, Elsevier. (c) The binding energy of x Li atoms adsorbed on PP (x=1, 8, 16, 32). Theoptimized stable configuration of Li_xP_{64} (x = 8, 16, 32) is shown as well. Reproduced with permission.[100] Copyright 2015, Royal Society of Chemistry. (d) Distance between the gas molecule and the phosphorene layer for the adsorption energy for CO, CO_2, NH_3, NO, and NO_2 on phosphorene. Reproduced with permission.[101] Copyright 2014, American Chemical Society. (e) Formation energies and relative formation energies of TM-doped M-BP systems with different defects and defect complexes (TM = Mn, Fe, and Co). Reproduced with permission.[112] Copyright 2016, American Chemical Society. (f) Comparison of adsorption energies of different adatoms on the P, BN, SiC, MoS_2, and graphene sheets. Reproduced with permission.[113] Copyright 2015, American Chemical Society.

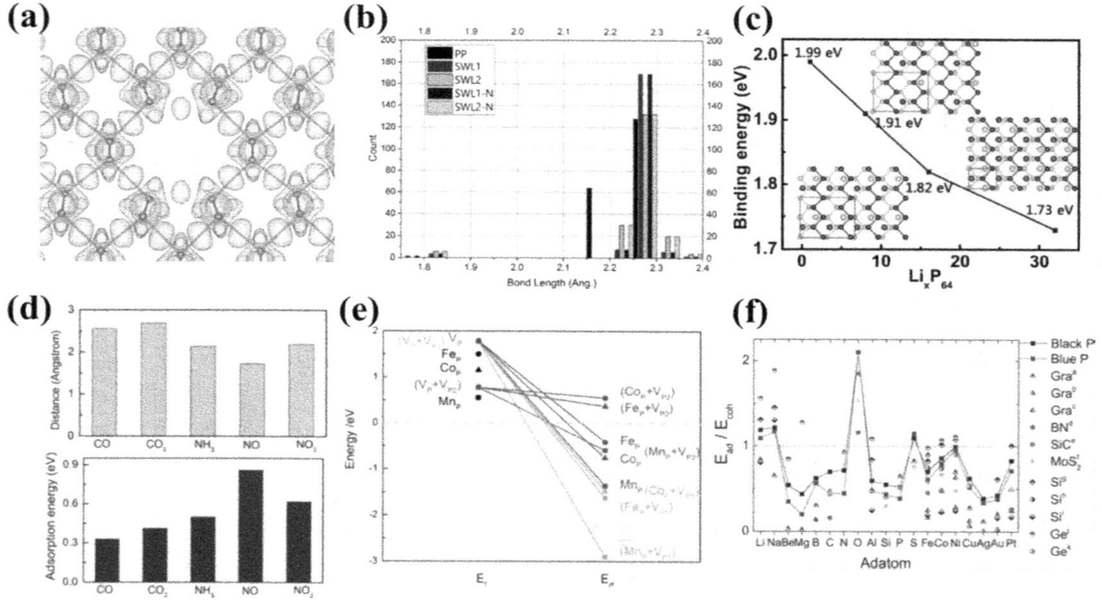

FIGURE 7.4 (a) Charge density distribution for point-defect phosphorene. Reproduced with permission.[99] Copyright 2020, Elsevier. (b) Bond length distribution for pristine, SW defected phosphorene, and N-doped SW defected phosphorene. Reproduced with permission.[41] Copyright 2018, Elsevier. (c) The binding energy of x Li atoms adsorbed on PP (x=1, 8, 16, 32). The optimized stable configuration of Li_xP_{64} (x = 8, 16, 32) is shown as well. Reproduced with permission.[100] Copyright 2015, Royal Society of Chemistry. (d) Distance between the gas molecule and the phosphorene layer for the adsorption energy for CO, CO_2, NH_3, NO, and NO_2 on phosphorene. Reproduced with permission.[101] Copyright 2014, American Chemical Society. (e) Formation energies and relative formation energies of TM-doped M-BP systems with different defects and defect complexes (TM = Mn, Fe, and Co). Reproduced with permission.[112] Copyright 2016, American Chemical Society. (f) Comparison of adsorption energies of different adatoms on the P, BN, SiC, MoS₂, and graphene sheets. Reproduced with permission.[113] Copyright 2015, American Chemical Society.

capability and lifetime of photogenerated carriers in the system, and affect the NRR activity and selectivity of BP.[106]

For BP doped with external impurities, when doped by non-metallic elements of the group IIIA and IVA, it will form a central undoped state; when doped by atoms of the group IIIA, it will transform into an indirect bandgap semiconductor, accompanied by enhanced optical absorption and strong anisotropy.[107] Marjaniet al. observed that Sc-doped BP could be transformed into a highly sensitive material for the detector of harmful hydrogen sulfide(H_2S) molecules.[108] It is well known that BP are easily oxidized in the environment, so Boukhvalov et al. and Doganov et al. revealed that the stability of BP can be effectively improved by fluorination, and that the BP passivated by graphene also shows enhanced transconductivity in the air.[109] It's mentioned that the BP adsorbed by metal atoms also effectively improved its stability.[110] The BP system adsorbed by Al atom would generate a temporal short-circuit current and open-circuit voltage under zero external bias,[111] while the BP system adsorbed by Ag atom showed high performance near infrared photodetection. In Figure 7.4(e), Wang et al. concluded that phosphoenes doped with transition metals V, Cr, Mn, Fe, or Ni could produce magnetism, while Co doping could not.[112] In addition, Ding et al. indicated that p-type and n-type dopants or intermediate gap states can be induced into phosphoene. Importantly, surface adsorption effectively functionalized phosphorus systems with a variety of spintronic properties. The comparison of different atomic adsorption energies is shown in Figure 7.4(f).[113]

Different from the defect BP the single-vacancy and double-vacancy arsenenes reduced the band gap,[114, 115] and the single-vacancy defects enhanced optical transition in the forbidden band region of arsenene, so the defect types could be identified by the absorption spectrum.[116] Zhou et al. and Yang et al. pointed out that the adsorption of HCN molecules on SW defective arsenene would change from the original physical adsorption to chemical adsorption,[117] and it showed high sensitivity and selectivity to SO_2.[118] Interestingly, according to Wang et al., when arsenene is doped with odd valence electrons, the system maintains its semiconducting properties, while when arsenene is doped with even valence electrons, the system becomes a metal.[119] Arsenene doped with transition metal atoms (V-, Fe-, Co-, and Ru) as NRR electrocatalyst has high stable adsorption potential of N_2 molecules.[120] Arsenene adsorbed by different metal atoms also exhibits rich properties including metal, semi-metal, semiconductor, and so on.[121] Depending on the degree of hydrogenation, the system is converted from a wide-band gap semiconductor to a metal and then to a Dirac metal material, from a semi-hydrogenated to a fully hydrogenated arsenene,[122] while Deobrat et al. accessed that arsenene adsorbed by organic molecules has an extensively reduced band gap.[123, 124]

Liu et al. concluded that the intrinsic defective antimonene has higher formation energy than arsenene, can induce fewer band gap states, and the Fermi level pinning was more suppressed.[125] The band gap of antimonene decreases with the decrease of SW defect concentration.[82] When antimonene is replaced by group VA atoms, local lattice deformation will be caused, resulting in a change of optoelectronic properties.[126] Yang and his team pointed out that antimonene adsorbed by metal atoms and some molecules shows the same rich physical properties as doped arsenene and can also be an eligible sensing material for polluted gases detection.[127, 128]

As a monoelemental material also from the group VA, bismuthene can be induced into local band gap state in the presence of vacancy defects, and the intrinsic point defects can regulate its optical, magnetic, and topological properties.[116] It is important to note that bismuthene also has a wealth of tunable electronic properties when doped with different TM atoms, such as semiconductors, semi-metals, and metals.[129] The most prominent feature is that the quantum spin Hall effect is achieved when the bismuthenes adsorb the homo-group and group VIA atoms, whereas the methyl-functionalized bismuthene will exhibit the 2D topological insulating edge state.[130–132]

7.3.4 GROUP VIA OF DEFECTIVE 2D XENES MATERIALS

The group VIA representatives of single element material, selenene and tellurene, have only been paid attention to in the last two years, with few related literature reports, and the study on their defect properties is still in the process of development. Here, only the excellent properties and applications of tellurene with intrinsic defects are given, and the related properties of selenene and tellurene with impurity defects need to be further explored.

In Figure 7.2(e), Xu et al. showed that when vacancy defects are formed in tellurene, the adsorption of the system for both O_2 and H_2O will be enhanced, and when vacancy and H_2O exist at the same time, the diffusion of O_2 will also be dramatically facilitated.[133] Another distinctive feature is that the vacancy defects can effectively regulate the band structure of tellurene, and the band gap will decrease with the increase of the vacancy concentration, and at the concentration of 5.3%, there will be an indirectransition to the direct band gap. Wang et al. have shown that the adsorption of nineteen typical adatoms (Li, Na, K, Ca, Fe,Co, Ni, Cu, Zn, Ag, Au, Pd, Pt, B, N, O, Si, Cl, and Al) and five typical gas molecules (H_2, O_2, H_2O, NO_2, and NH_3) on tellurene (Figure 7.2(f)).[134] Most of these adsorbed atoms are chemically adsorbed on tellurene, resulting in structural deformation and local recombination. Xu et al. further studied the application value of tellurene with vacancy defects. SW defects and single vacancy defects can promote the sensitivity detection of tellurene on the adsorption strength and electronic transport performance of nitrogen oxide molecules, which are a great candidate material for the family of nitrogen oxide molecules gas sensors, as shown in Figure 7.2(d).[135]

7.4 OPTOELECTRONIC APPLICATIONS OF DEFECTIVE XENES

Due to the fact that the application of 2D monoelemental materials is mostly focused on the BP structure, we briefly summarize the possible optoelectronic applications of 2D defective Xenes materials by taking the defective BP as an example.

FETs is a kind of semiconductor device that uses the electric field effect as the input circuit to control the current in the output circuit. Ryder et al. have shown that covalently bonded and aryldiazo functionalized BP FETs also have excellent mobility and on-off current ratios and are relatively stable in the environment, as shown in the Figure 7.5(a).[136] A photodetector is a sensor that can convert an optical signal into an electrical signal. Kang et al. respectively reported the preparation of n- and p-doped BP photodetector using 3-amino-propyl-triethoxysilane (APTES) and octadecyltrichlorosilane (OTS) as shown in Figure 7.5(a),[136] which have significantly improved responsivity and excellent photore-sponse. And the gas sensor is a kind of gas volume fraction can be converted

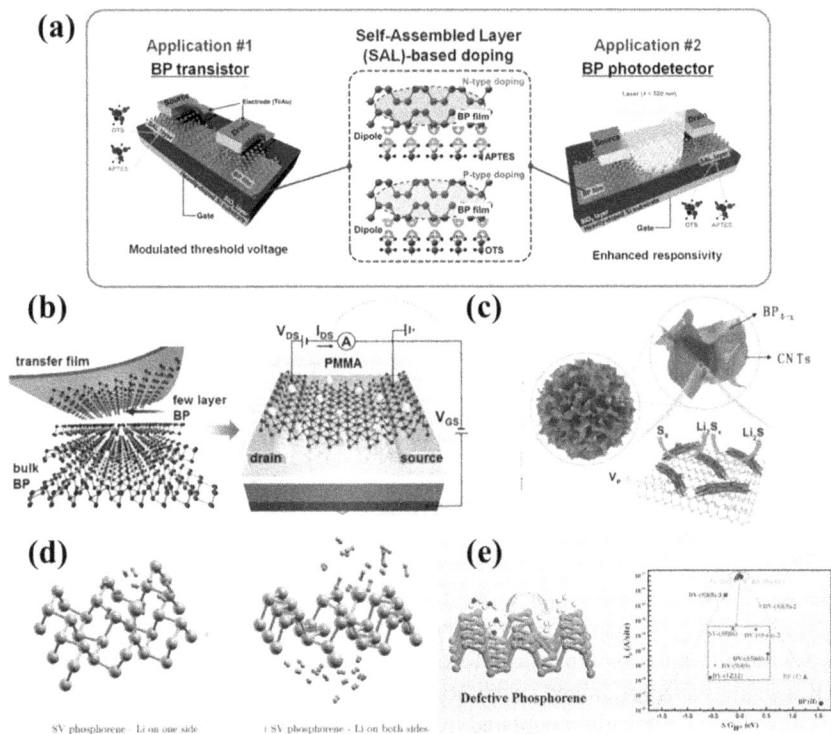

FIGURE 7.5 (a) Schematic illustration/optical image of BP transistor devices edited by 3-amino-propyltriethoxysilane (APTES) and octadecyltrichlorosilane (OTS) on the left and SAL-doped BP photodetector under laser illumination on the right. Reproduced with permission.[136] Copyright 2017, American Chemical Society. (b) Schematic of the fabrication of Pt-functionalized BP hydrogen sensors. Reproduced with permission.[137] Copyright 2017, American Institute of Physics. (c) Schematic diagram of the functional-ities of BP_{4-x}/carbon nanotubes. Reproduced with permission.[138] Copyright 2021, Elsevier. (d) Monovacancy phosphorene for H_2 storage with Li decoration on one side and both sides. Reproduced with permission.[139] Copyright 2020, Elsevier. (e) Schematic diagram of catalytic reaction process and exchange current density of Pt (111), P-phosphorene, and defective phosphorene. Reproduced with permission.[140] Copyright 2021, Elsevier.

into the corresponding electrical signal converter. Lee et al., by exfoliating BP flakes with Pt nanoparticles effectively improved the sensitivity and efficiency of Pt-functionalized BP sensors as hydrogen sensors (Figure 7.5(b)).[137] A battery is a device which can change chemical energy into electrical energy. Atashzar et al. predicted that the existence of point defects in the BP system significantly enhanced the binding energy between the alkali metal and the original BP and reduced the diffusion barrier. The battery prepared, based on the BP defect, had excellent electrode material properties, such as fast migration rate, fast charge/discharge capability, high theoretical storage capacity, low average open-circuit voltages, and so on. Xu et al. discovered that defective phosphorene-based sulfur cathodes can be used for high-efficiency lithium-sulfur batteries as shown in (Figure 7.5(c)).[138] Hydrogen as a clean energy has become the demand of human life. Hydrogen storage material is a kind of material that can absorb and release hydrogen reversibly. Haldar et al. conducted a detailed study of Li-modified vacancy BP, which significantly improved the adsorption and storage capacity of hydrogen, and the gravimetric density of hydrogen storage reached about 5.3%, as shown in Figure 7.5(d).[139]

In the catalytic reaction, the reaction path can be changed by the chemical interaction between the catalyst and the reactants, thus reducing the activation energy of the reaction. In (Figure 7.5(e)), Wang et al. revealed that defect engineering broke the inertness of the BP surface and added more adsorption sites.[140] Therefore, BP with vacancy defect showed better catalytic hydrogen evolution performance than P, and its energy barrier for the Tafel reaction in the H_2 desorption process was effectively reduced. Of course, in addition to the applications listed here, BP has great potential in many other applications.[53, 141–143]

7.5 SUMMARY AND PERSPECTIVES

This chapter gives a comprehensive review of defects in different modes of mono-element 2D materials from the group IIIA to group VIA, as well as the structures of systems after defect engineering and related properties. First of all, the defect types include common intrinsic defects and defects introduced by external impurities such as adsorption and substitution. These defective single-element materials have been effectively modulated in terms of energy band structure, electronic properties, and optical properties, etc. Therefore, their application potential in field effect transistors, photodetectors, sensors, batteries, hydrogen storage, and catalysis has been significantly enhanced.

Despite the many advantages of defective monoelemental materials, there are still many problems and challenges. On the one hand, at present, research on the structural properties and applications of defective monoelemental materials is confined to theoretical prediction, while experimental preparation and practical application need to be kept up to date. First of all, as far as the intrinsically single-element 2D materials are concerned, it is difficult to be prepared in a large area, let alone applied in practice. Secondly, it is difficult to achieve the defect types and doping sites required by the theory in an accurate and controllable way in experiments, so as to ensure the optimization of properties. Moreover, the theoretical research on the defective monoelemental materials is not comprehensive and rich, and the properties of many group IVA and group VIA Xenes materials after defects are still unknown. Another big aspect is that single-element 2D materials generally have the problem of environmental stability. Although the stability can be improved to some extent through defect engineering, the defect structure itself is not very stable. How to guarantee the sustainable stability and apply it to the process is another big problem.

Once the above mentioned problems are overcome, defective monoelemental materials will become high-quality materials for the emerging technology industry and accelerate the development of science and technology. On this basis, the potential applications of defective Xenes materials in flexible devices, memory storage devices, spintronic devices and biomedical nanodevices can be further expanded. At the same time, it can also combine defect monoelemental materials with other high-quality 2D materials to form heterojunctions, exert other external field effects on the basis of defects, realize more abundant defect types through different means, and further expand the market and space of single-element 2D materials by combining multiple means.

REFERENCES

1. Geim, A. K. & Novoselov, K. S., The rise of graphene. *Nature Materials*, 6, 9, 2007.
2. Geim, A. K., Graphene: Status and prospects. *Science*, 324, 5934, 2009.
3. Balandin, A. A. et al., Superior thermal conductivity of single-layer graphene. *Nano Letters*, 8, 3, 2008.
4. Liu, Y., Ge, X. & Li, J., Graphene lubrication. *Applied Materials Today*, 20, 100662, 2020.
5. Stoller, M., Park, S., Zhu, Y., An, J. & Ruoff, R. S., Graphene-based ultracapacitors. *Nano Letters*, 8, 10, 2008.
6. Novoselov, K. S. et al., Two-dimensional gas of Massless Dirac Fermions in Graphene. *Nature*, 438, 7065, 2005.
7. Ferrari, A. C. et al., Raman spectrum of graphene and graphene layers. *Physical Review Letters*, 97, 18, 2006.
8. Wang, Y. et al., Nitrogen-doped graphene and its application in electrochemical biosensing. *Acs Nano*, 4, 1790, 2010.
9. Sabio, J., Sols, F. & Guinea, F., Two-body problem in graphene. *Physical Review. B, Condensed Matter*, 81, 4, 2009.
10. Gui et al., Band structure engineering of graphene by strain: First-principles calculations. *Physical Review B*, 78, 7, 2008.
11. Jia, X., Campos-Delgado, J., Terrones, M., Meunier, V. & Dresselhaus, M. S., Graphene edges: A review of their fabrication and characterization. *Nanoscale*, 3, 1, 2011.
12. Xiong, G., Meng, C., Reifenberger, R. G., Irazoqui, P. P. & Fisher, T. S., A review of Graphene-based electrochemical microsupercapacitors. *Electroanalysis*, 26, 1, 2014.
13. Ni, Z. H. et al., Uniaxial strain on graphene: Raman spectroscopy study and band-gap opening. *Acs Nano*, 2, 11, 2008.
14. Dvorak, M., Oswald, W. & Wu, Z., Bandgap opening by patterning Graphene. *Scientific Reports*, 3, 1, 2013.

15. Zhou, S. Y. et al., Substrate-induced band gap opening in epitaxial graphene. *Nature Materials*, 6, 10, 2007.
16. A, H. T. et al., Growth of epitaxial graphene: Theory and experiment. *Physics Reports*, 542, 3, 2014.
17. Schwierz, F., Graphene transistors. *Nature Nanotechnology*, 5, 7, 2010.
18. Guo, B., Fang, L., Zhang, B. & Gong, J. R., Graphene doping: A review. *Insciences Journal*, 1, 2, 2011.
19. Kislenko, V. A., Pavlov, S. V. & Kislenko, S. A., Influence of defects in graphene on electron transfer kinetics: The role of the surface electronic structure. *Electrochimica Acta*, 341, 136011, 2020.
20. Chen, J. H. et al., Defect scattering in graphene. *Physical Review Letters*, 102, 23, 2009.
21. Bozkurt, P., Oleynik, I. I., Batzill, M., Lahiri, J. & Lin, Y., An extended defect in graphene as a metallic wire. *Nature Nanotechnology*, 5, 5, 2010.
22. Meng, C. et al., Restoration of graphene from graphene oxide by defect repair. *Carbon*, 50, 7, 2012.
23. Wu, T. et al., Nitrogen and boron doped monolayer graphene by chemical vapor deposition using polystyrene, urea and boric acid. *New Journal of Chemistry*, 36, 6, 2012.
24. Jia, Y. et al., Defect graphene as a trifunctional catalyst for electrochemical reactions. *Advanced Materials*, 28, 43, 2016.
25. Banhart, F., Kotakoski, J. & Krasheninnikov, A. V., Structural defects in graphene. *Acs Nano*, 5, 1, 2011.
26. Lusk, M. T. & Carr, L. D., Nano-engineering defect structures on graphene. *Physical Review Letters*, 100, 17, 2007.
27. Coleman & Jonathan, N., Liquid exfoliation of defect-free graphene. *Accounts of Chemical Research*, 46, 1, 2013.
28. Jain, A. & Mcgaughey, A., Strongly anisotropic in-plane thermal transport in single-layer black phosphorene. *Scientific Reports*, 110, 115, 2015.
29. Zh, A. et al., Structures, properties and application of 2D monoelemental materials (Xenes) as graphene analogues under defect engineering - ScienceDirect. *Nano Today*, 35, 100906, 2020.
30. Huang, M. et al., Broadband black-Phosphorus photodetectors with high responsivity. *Advanced Materials*, 28, 18, 2016.
31. Lherbier, A., Botello-Méndez, A. R. & Charlier, J.-C., Electronic and optical properties of pristine and oxidized borophene. *2D Materials*, 3, 4, 2016.
32. Li, J. Y. et al., Tuning the electronic and magnetic properties of borophene by 3d transition-metal atom adsorption. *Physics Letters A*, 380, 46, 2016.
33. Huang, M. et al., Broadband Black-Phosphorus photodetectors with high responsivity. *Advanced Materials*, 28, 18, 2016.
34. Zhang, S. et al., Recent progress in 2D group VA semiconductors: from theory to experiment. *Chemical Society Reviews*, 47, 3, 2018.
35. Doganov, R.A. et al., Transport properties of pristine few-layer black phosphorus by van der Waals passivation in an inert atmosphere. *Nature Communication*, 6, 6647, 2015.
36. Hu, W. & Yang, J., Defects in Phosphorene. *The Journal of Physical Chemistry C*, 119, 35, 2015.
37. Gablech, I. et al., Monoelemental 2D materials-based field effect transistors for sensing and biosensing: Phosphorene, antimonene, arsenene, silicene, and germanene go beyond graphene. *TRAC Trends in Analytical Chemistry*, 105, 251, 2018.

38. Boukhvalov, D. W., Rudenko, A. N., Prishchenko, D. A., Mazurenko, V. G. & Katsnelson, M. I., Chemical modifications and stability of phosphorene with impurities: a first principles study. *Physical Chemistry Chemical Physics*, 17, 23, 2015.

39. Guo, Y. & Robertson, J., Vacancy and doping states in monolayer and bulk Black Phosphorus. *Scientific Reports*, 5, 14165, 2015.

40. Kistanov, A. A. et al., Exploring the charge localization and band gap opening of borophene: a first-principles study. *Nanoscale*, 10, 3, 2018.

41. Rezaee, A.E., Almasi Kashi, M. & Baktash, A., Stone-Wales like defects formation, stability and reactivity in black phosphorene. *Materials Science and Engineering: B*, 236-237, 2018.

42. Liu, Y., Xu, F., Zhang, Z., Penev, E. S. & Yakobson, B. I., Two-dimensional monoelemental semiconductor with electronically inactive defects: the case of phosphorus. *Nano Letters*, 14, 12, 2014.

43. Xia, W., Hu, W., Li, Z. & Yang, J., A first-principles study of gas adsorption on germanene. *Physical Chemistry Chemical Physics*, 16, 41, 2014.

44. Üzengi Aktürk, O., Aktürk, E. & Ciraci, S., Effects of adatoms and physisorbed molecules on the physical properties of antimonene. *Physical Review B*, 93, 3, 2016.

45. Navid, I. A. & Subrina, S., Thermal transport characterization of carbon and silicon doped stanene nanoribbon: an equilibrium molecular dynamics study. *Rsc Advances*, 8, 55, 2018.

46. Zhang, R., Waters, J., Geim, A. K. & Grigorieva, I. V., Intercalant-independent transition temperature in superconducting black phosphorus. *Nature Communications*, 8, 1, 2017.

47. Walia, S. et al., Ambient protection of few-layer black phosphorus via sequestration of reactive oxygen species. *Advanced Materials*, 29, 27, 2017.

48. Zhu, X. et al., Stabilizing black phosphorus nanosheets via edge-selective bonding of sacrificial C60 molecules. *Nature Communications*, 9, 1, 2018.

49. Chen, X. et al., High quality sandwiched black phosphorus heterostructure and its quantum oscillations. *Nature Communications*, 6, 7315, 2015.

50. Li, Z. & Chen, F., Ion beam modification of two-dimensional materials: Characterization, properties, and applications. *Applied Physics Reviews*, 4, 1, 2017.

51. Tianbing, S. et al., Creating an air-stable Sulfur-Doped Black Phosphorus-TiO_2 composite as high-performance Anode material for Sodium-Ion storage. *Advanced Functional Materials*, 29, 22, 2019.

52. Liu et al., GeH: A novel material as a visible-light driven photocatalyst for hydrogen evolution. *Chemical Communications- Royal Society of Chemistry*, 50, 75, 2014.

53. Ryder, C. R. et al., Covalent functionalization and passivation of exfoliated black phosphorus via aryl diazonium chemistry. *Nature Chemistry*, 8, 6, 2016.

54. Cultrara, N. D. et al., Synthesis of 1T, 2H and 6R Germanane Polytypes. *Chemistry of Materials*, 30, 4, 2018.

55. Kistanov, A. et al., Exploring the charge localization and band gap opening of Borophene: A first-principles study. *Nanoscale*, 10, 3, 2018.

56. Yang et al., Borophene as an extremely high capacity electrode material for Li-ion and Na-ion batteries. *Nanoscale*, 8, 597, 2016.

57. Zhang, Z., Yang, Y., Penev, E. S. & Yakobson, B. I., Elasticity, flexibility and ideal strength of Borophenes. *Advanced Functional Materials*, 27, 9, 2016.

58. Arabha, S., Akbarzadeh, A. H. & Rajabpour, A., Engineered porous borophene with tunable anisotropic properties. *Composites Part B: Engineering*, 200, 108260, 2020.

59. Yu, X., Chen, F., Yu, Z. & Li, Y., Computational study of Borophene with line defects as sensors for Nitrogen-containing gas molecules. *ACS Applied Nano Materials*, 3, 10, 2020.

60. Chowdhury, S., Majumdar, A. & Jana, D., Electronic and optical properties of the supercell of 8-Pmmn borophene modified on doping by H, Li, Be, and C: A DFT approach. *Applied Physics A*, 125, 5, 2019.

61. Wang, L. et al., First-principles investigation on hydrogen storage performance of Li, Na and K decorated borophene. *Applied Surface Science*, 427, 1030, 2017.

62. Baraiya, B. A. et al., Nitrogen-decorated borophene: An empowering contestant for hydrogen storage. *Applied Surface Science*, 527, 146852, 2020.

63. Tu, X. et al., First-principles study of pristine and Li-doped borophene as a candidate to detect and scavenge SO2gas. *Nanotechnology*, 32, 32, 2021.

64. Lherbier, A., Botello-Méndez, A. & Charlier, J. C., Electronic and optical properties of pristine and oxidized borophene. *2D Materials*, 3, 4, 2016.

65. Zheng, F. B. & Zhang, C. W., The electronic and magnetic properties of functionalized silicene: A first-principles study. *Nanoscale Research Letters*, 7, 1, 2012.

66. Kulish, V. V., Surface reactivity and vacancy defects in single-layer borophene polymorphs. *Physical Chemistry Chemical Physics*, 19, 18, 2017.

67. Özçelik, V. O., Gurel, H. H. & Ciraci, S., Self-healing of vacancy defects in single-layer graphene and silicene. *Physical Review B*, 88, 4, 2013.

68. Gao, J., Zhang, J., Liu, H., Zhang, Q. & Zhao, J., Structures, mobilities, electronic and magnetic properties of point defects in silicene. *Nanoscale*, 5, 20, 2013.

69. Shojaeverdi, B. & Zaminpayma, E., Influence of vacancy cluster on the electronic transport properties of silicene sheet. *Physica E: Low-dimensional Systems and Nanostructures*, 121, 114109, 2020.

70. Kamyabmehr, S., Zoriasatain, S. & Farhang Matin, L., Effects of Stone-Wales defects on optical properties of silicene: DFT study. *Optik*, 241, 166952, 2021.

71. Sahin, H., Sivek, J., Li, S., Partoens, B. & Peeters, F. M., Stone-Wales defects in silicene: Formation, stability, and reactivity of defect sites. *Physical Review B*, 88, 4, 2013.

72. Hu, W., Wu, X., Li, Z. & Yang, J., Porous silicene as a hydrogen purification membrane. *Physical Chemistry Chemical Physics*, 15, 16, 2013.

73. Hussain, T., Kaewmaraya, T., Chakraborty, S. & Ahuja, R., Defect and substitution induced silicene sensor to probe toxic gases. *The Journal of Physical Chemistry C*, 120, 44, 2016.

74. Chandiramouli, R., Srivastava, A. & Nagarajan, V., First-principles insights of CO adsorption characteristics on Ge and In substituted silicene nanosheet. *Silicon*, Accepted, 3, 2017.

75. Pablo-Pedro, R. et al., Understanding disorder in 2D materials: The case of Carbon doping of Silicene. *Nano Letters*, 20, 9, 2020.

76. Azevedo, S. & Kaschny, J. R. Hydrogenated BN monolayers: A first principles study. *European Physical Journal B*, 86, 9, 2013.

77. Das, R., Chowdhury, S., Majumdar, A. & Jana, D., Optical properties of P and Al doped silicene: a first principles study. *Rsc Advances*, 5, 1, 2015.

78. Zhang, P., Li, X. D., Hu, C. H., Wu, S. Q. & Zhu, Z. Z., First-principles studies of the hydrogenation effects in silicene sheets. *Physics Letters A*, 376, 14, 2012.

79. Yao, Y., Liu, G. & Yang, J., Effect of shear deformation on aluminum adsorption on silicene. *Journal of Molecular Structure*, 1234, 130172, 2021.

80. Hussain, T., Kaewmaraya, T., Chakraborty, S. & Ahuja, R., Defect and substitution-induced Silicene sensor to probe toxic gases. *The Journal of Physical Chemistry C*, 120, 44, 2016.

81. Zhu, L. et al., Defective germanene as a high-efficiency helium separation membrane: a first-principles study. *Nanotechnology*, 28, 13, 2017.
82. Shen, L., Lan, M., Zhang, X. & Xiang, G., The structures and diffusion behaviors of point defects and their influences on the electronic properties of 2D stanene. *Rsc Advances*, 7, 16, 2017.
83. Xu, Q. et al., First-principles calculation of optimizing the performance of Germanene-based supercapacitors by vacancies and metal atoms. *The Journal of Physical Chemistry C*, 124, 23, 2020.
84. Wang, Y. & Ding, Y., Electronic structure and carrier mobilities of Arsenene and Antimonene Nanoribbons: A first-principle study. *Nanoscale research letters*, 10, 1, 2015.
85. Xiao, P., Fan, X. L. & Liu, L. M., Tuning the electronic properties of half- and full-hydrogenated germanene by chlorination and hydroxylation: A first-principles study. *Computational Materials Science*, 92, 244, 2014.
86. Li et al., Tunable electronic and magnetic properties in germanene by alkali, alkaline-earth, group III and 3d transition metal atom adsorption. *Physical Chemistry Chemical Physics Cambridge Royal Society of Chemistry*, 118, 43, 2014.
87. Zhang, J. M., Wang, S. F., Chen, L.Y., Xu, K. W. & Ji, V., Structural, electronic and magnetic properties of the 3d transition metal atoms adsorbed on boron nitride nanotubes. *The European Physical Journal B*, 76, 2, 2010.
88. Yang, Y., Zhang, H., Song, L. & Liu, Z., RETRACTED: Adsorption of gas molecules on the defective stanene nanosheets with single vacancy: A DFT study. *Applied Surface Science*, 512, 145727, 2020.
89. Abbasi, A. & Sardroodi, J. J., Electronic structure tuning of stanene monolayers from DFT calculations: effects of substitutional elemental doping. *Applied Surface Science*, 456, 290, 2018.
90. Hongyan, L., Guixian, T., Bin, H., Dan, L. & Liu, Z., RETRACTED: Efficient band gap opening in single-layer stanene via patterned Ga-As codoping: Towards semiconducting nanoelectronic devices. *Synthetic Metals*, 264, 116388, 2020.
91. Aaab, C., DFT study of the effects of Al P pair doping on the structural and electronic properties of stanene nanosheets. *Physica E: Low-dimensional Systems and Nanostructures*, 108, 34, 2019.
92. Wang, D., Gao, H., Xiang, Y. & Jiang, L., Tuning the structural and electronic properties of single-layer stanene through doping 4d transition metals (Mo, Nb, Rh and Ru): A DFT study. *Synthetic Metals*, 264, 116399, 2020.
93. Tan, Y., Xu, Y. & Ao, Z., Nitrogen fixation on a single Mo atom embedded stanene monolayer: a computational study. *Physical Chemistry Chemical Physics*, 22, 25, 2020.
94. Kadioglu et al., Adsorption of alkali and alkaline-earth metal atoms on stanene: A first-principles study. *Materials Chemistry & Physics*, 180, 326, 2016.
95. Hashemi, D. & Iizuka, H., Magnetic properties of 3d transition metal (Sc–Ni) doped plumbene. *Rsc Advances*, 10, 12, 2020.
96. Yuan, S., Rudenko, A. N. & Katsnelson, M. I., Transport and optical properties of single- and bilayer black phosphorus with defects. *Physical Review B*, 91, 11, 2015.
97. Rivero, P. et al., Simulated scanning tunneling microscopy images of few-layer-phosphorus capped by graphene and hexagonal boron nitride monolayers. *Physical Review B*, 91, 11, 2014.
98. Liu, Y., Xu, F., Zhang, Z., Penev, E. S. & Yakobson, B. I., Two-dimensional monoelemental semiconductor with electronically inactive defects: The case of phosphorus. *Nano Letters*, 14, 12, 2012.

99. Rezaee, A. E. & Almasi Kashi, M., The influence of point defects on Na diffusion in black phosphorene: First principles study. *Journal of Physics and Chemistry of Solids*, 143, 109432, 2020.

100. Guo, G. C., Wei, X. L., Wang, D., Luo, Y. & Liu, L. M., Pristine and defect-containing phosphorene as promising anode materials for rechargeable Li batteries. *Journal of Materials Chemistry A*, 3, 21, 2015.

101. Kou, L., Frauenheim, T. & Chen, C., Phosphorene as a superior gas sensor: Selective adsorption and distinct I-V response. *Journal of Physical Chemistry Letters*, 5, 15, 2014.

102. Ghashghaee, M. & Ghambarian, M., Highly improved carbon dioxide sensitivity and selectivity of black phosphorene sensor by vacancy doping: A quantum chemical perspective. *International Journal of Quantum Chemistry*, 120, 16, 2020.

103. Ghambarian, M., Azizi, Z. & Ghashghaee, M., Remarkable improvement in Phosgene detection with defect-engineered Phosphorene sensor: First-principles calculations. *Physical Chemistry Chemical Physics*, 22, 17, 2020.

104. Ghadiri, M., Ghashghaee, M. & Ghambarian, M., Defective phosphorene for highly efficient formaldehyde detection: Periodic density functional calculations. *Physics Letters A*, 384, 31, 2020.

105. Ghadiri, M., Ghambarian, M. & Ghashghaee, M., Detection of CNX cyanogen halides (X = F, Cl) on metal-free defective phosphorene sensor: Periodic DFT calculations. *Molecular Physics*, 119, 4, 2020.

106. Pei, W., Zhou, S., Zhao, J., Du, Y. & Dou, S. X., Optimization of photocarrier dynamics and activity in phosphorene with intrinsic defects for nitrogen fixation. *Journal of Materials Chemistry A*, 8, 39, 2020.

107. Guo, Y. & Robertson, J., Vacancy and doping states in monolayer and bulk black phosphorus. *Scientific Reports*, 5, 1, 2015.

108. Marjani, A., Ghashghaee, M., Ghambarian, M. & Ghadiri, M., Scandium doping of black phosphorene for enhanced sensitivity to hydrogen sulfide: Periodic DFT calculations. *Journal of Physics and Chemistry of Solids*, 148, 109765, 2021.

109. Doganov, R. A. et al., Transport properties of pristine few-layer black phosphorus by van der Waals passivation in an inert atmosphere. *Nature Communications*, 6, 6647, 2015.

110. Guo, Z. et al., Metal-ion-modified black phosphorus with enhanced stability and transistor performance. *Advanced Materials*, 29, 42, 2017.

111. Liu, Y., Cai, Y., Zhang, G., Zhang, Y. W. & Ang, K. W., Al-Doped Black Phosphorus p–n Homojunction Diode for high performance photovoltaic. *Advanced Functional Materials*, 27, 7, 2017.

112. Wang, Y., Pham, A., Li, S. & Yi, J., Electronic and magnetic properties of transition-metal-doped monolayer black phosphorus by defect engineering. *The Journal of Physical Chemistry C*, 120, 18, 2016.

113. Ding, Y. & Wang, Y., Structural, electronic, and magnetic properties of adatom adsorptions on Black and Blue Phosphorene: A first-principles study. *The Journal of Physical Chemistry C*, 119, 19, 2015.

114. Liang et al., Characterization of point defects in monolayer arsenene. *Applied Surface Science: A Journal Devoted to the Properties of Interfaces in Relation to the Synthesis and Behaviour of Materials*, 443, 74, 2018.

115. Sun, X. et al., Structures, mobility and electronic properties of point defects in arsenene, antimonene and an antimony arsenide alloy. *Journal of Materials Chemistry C*, 5, 17, 2017.

116. Kadioglu, Y. et al., Modification of electronic, magnetic structure and topological phase of bismuthene by point defects. *Physical Review B*, 96, 24, 2017.

117. Zhou, Q., Ju, W., Liu, Y., Li, J. & Zhang, Q., Influence of defects and dopants on the sensitivity of arsenene towards HCN. *Applied Surface Science*, 506, 144936, 2020.
118. Yang, A. et al., Tunable SO2-sensing performance of arsenene induced by Stone-Wales defects and external electric field. *Applied Surface Science*, 523, 146403, 2020.
119. Wang, Y. P., Zhang, C. W., Ji, W. X. & Wang, P. J., Unexpected band structure and half-metal in non-metal-doped arsenene sheet. Applied Physics Express, 8, 6, 2015.
120. Xu, Z. et al., Single atom-doped arsenene as electrocatalyst for reducing nitrogen to ammonia: A DFT study. *Physical Chemistry Chemical Physics*, 22, 45, 2020.
121. Ni et al., The realization of half-metal and spin-semiconductor for metal adatoms on arsenene. *Applied Surface Science A Journal Devoted to the Properties of Interfaces in Relation to the Synthesis & Behaviour of Materials*, 390, 60, 2016.
122. Cai et al., First principles investigation of small molecules adsorption on Antimonene. *IEEE Electron Device Letters*, 38, 1, 2017.
123. Yang et al., Light adatoms influences on electronic structures of the two-dimensional arsenene nanosheets, 230, 6, 2016.
124. Singh D, Gupta S K, Sonvane Y, et al. Modulating the electronic and optical properties of monolayer arsenene phases by organic molecular doping. *Nanotechnology*, 28, 49, 2017.
125. Liu, Y. et al., Band structure, band offsets, and intrinsic defect properties of few-layer Arsenic and Antimony. *The Journal of Physical Chemistry C*, 124, 13, 2020.
126. Hu, Y. et al., Arsenene and antimonene doped by group VA atoms: First-principles studies of the geometric structures, electronic properties and STM images. *Physica B: Condensed Matter*, 553, 159, 2018.
127. Kistanov, A. et al., A first-principles study on the adsorption of small molecules on Antimonene: Oxidation tendency and stability. *Journal of Materials Chemistry C*, 6, 15, 2018.
128. Ahin, H., Cahangirov, S., Topsakal, M., Bekaroglu, E. & Ciraci, S., Monolayer honeycomb structures of group IV elements and III-V binary compounds: First-principles calculations. *Physical Review B*, 80, 15, 2009.
129. Mengyu et al., Prediction of electronic and magnetic properties in 3d-transition-metal X-doped bismuthene (X = V, Cr, Mn and Fe). *Applied Surface Science*, 486, 2019.
130. Liu, C. C. et al., Low-energy effective Hamiltonian for giant-gap quantum spin hall insulators in honeycomb X-Hydride/Halide (X=N-Bi) monolayers. *Physical. Review.B*, 90, 8, 2014.
131. Niu, C. et al., Functionalized bismuth films: Giant gap quantum spin Hall and valley-polarized quantum anomalous Hall states. *Physical Review B*, 91, 4, 2015.
132. Liu, M. Y., Chen, Q., Cao, C. & He, Y., Realization of versatile electronic, magnetic properties and new topological phases in hydrogenated bismuthene. *Electronic Structure*, 1, 2, 2019.
133. Xu, W., Gan, L., Wang, R., Wu, X. & Xu, H., Surface adsorption and vacancy in tuning the properties of Tellurene. ACS *Applied Materials & Interfaces*, 12, 16, 2020.
134. Wang, X. et al., Effects of adatom and gas molecule adsorption on the physical properties of tellurene: a first principles investigation. *Physical Chemistry Chemical Physics*, 20, 6, 2018.
135. Xu, Z. et al., Gas sensing properties of defective tellurene on the nitrogen oxides: A first-principles study. *Sensors and Actuators A: Physical*, 328, 112766, 2021.
136. Kang, D.-H. et al., Self-Assembled Layer (SAL)-based doping on Black Phosphorus (BP) transistor and photodetector. *ACS Photonics*, 4, 7, 2017.

137. Lee, G., Jung, S., Jang, S., et al. Platinum-functionalized black phosphorus hydrogen sensors. *Applied Physics Letters*, 110, 242103, 2017.

138. Xu, Y, Zhang, W, Ma, H., et al. Engineering the 3D framework of defective Phosphorene-based Sulfur Cathodes for High-efficiency Lithium-Sulfur batteries. *Electrochimica Acta*, 392, 139025, 2021.

139. Haldar, S., Mukherjee, S., Ahmed, F. & Singh, C. V., A first principles study of hydrogen storage in lithium decorated defective phosphorene. *International Journal of Hydrogen Energy*, 42, 36, 2017.

140. Wang, M. et al., Defects engineering promotes the electrochemical hydrogen evolution reaction property of phosphorene surface. *International Journal of Hydrogen Energy*, 46, 2, 2021.

141. Shishi et al., Improving the stability and optical properties of germanane via one-step covalent methyl-termination. *Nature Communications*, 5, 1, 2014.

142. Liu, B. et al., Black Arsenic-Phosphorus: Layered Anisotropic infrared semiconductors with highly tunable compositions and properties. *Advanced Materials*, 27, 30, 2015.

143. Tang, X. et al., Quantum Dots: Fluorination-enhanced ambient stability and electronic tolerance of Black Phosphorus Quantum Dots. *Advanced Science*, 5, 9, 2018.

8 2D Xenes Materials under Different Field Actions

Yujie Liao and Zongyu Huang

CONTENTS

8.1 INTRODUCTION

Two-dimensional (2D) materials, such as graphene, are widely used systems with many excellent properties. And encouraged by the successful synthesis of graphene, a plethora of 2D materials have been successfully prepared and applied to various property-related applications, which greatly promoted the rise and development of 2D materials.[1–9] Recently, 2D monoelemental materials (Xenes) represented by phosphorene, also as graphene analogous nanomaterials, have rapidly become members of the 2D material family that are attracting great attention. As a novel single-element two-dimensional material, it has been reported that it is mainly distributed in the periodic table of IIIA (B, Al, Ga, In), IVA (Si, Ge, Sn, Pd), VA (P, As, Sb, and Bi) and VIA (Se and Te).[10, 11]

Being identical to the traditional 2D materials, the 2D Xenes also show remarkable properties and wide applications in the fields of electronics and optics. They can exist in different forms, such as metal, semi-metal, and semi-conductor, and have adjustable physical properties. 2D Xenes have a number of advantages such as the characteristics of ultrahigh specific surface area, ultrathin thickness, high carrier mobility, tunable band gap, and in-plane anisotropy. Compared to h-BN (wide band gap) and graphene (zero band gap), the band gap of Xenes can be adjusted in a variety of ways, which is conducive to optical sensing applications. Furthermore, due to their high specific surface area, simple element composition, and good biodegradability, Xenes have great prospects in the biomedical field. Xenes are attractive in the field of polarized

DOI: 10.1201/9781003207122-8

photodetection and neural networks due to in-plane anisotropy caused by the fold structure. Theoretical and experimental investigations have confirmed that the Xenes endow variable band structure with tunable physical properties from metal to semiconductor and insulators. Considering their tunable physical properties, mechanical robustness and flexibility, diverse surface/edge terminations, Xenes have been suggested to be potential candidates for future nanodevices. Therefore, the Xenes are found to be versatile and have the potential to explore a variety of new applications.[12–21]

However, the mechanical and electrical properties of 2D Xenes are poorer than those of graphene. The band gap of pristine Xenes is generally not sufficiently large for their application as semiconductors. Moreover, a robust magnetism, which is crucial for applications in spintronics, has not been found. In summary, Xenes require further modification to broaden their applications in more fields. With the rapid development in recent years, various studies have shown that the properties of 2D materials are often improved under the action of external fields, such as strain, electric field, and magnetic field. In theory, the strategies of strain, electric field, and magnetic field can also effectively control the basic properties of 2D Xenes, thus providing a broad prospect for the practical application of these materials. Herein, we summarize the modulation effects of the outfield on 2D Xenes and provide the basis for their application and development.

8.2 MODULATION OF PROPERTIES IN 2D MATERIALS

To expand the applications of graphene, researchers are exploring ways to open its band gaps. Inspired by this idea and expecting to amplify the advantages and make up for the disadvantages of 2D materials, people began to study regulating the electronic structure of two-dimensional materials to obtain better physical and chemical properties. Below, the common modulate project of properties in 2D materials are briefly introduced.

Strain: the application of mechanical strain will change the crystal lattice constant and reduce the crystal symmetry, thus leading to significant changes in the band edge, which is usually accompanied by the decay and splitting of energy levels, the corresponding change of effective mass, and the change of related properties such as electronic properties and optical properties. In the experiment, we often encounter the heterojunction system containing strain, the stress problem in the growth process of the film, and the change of material properties caused by the introduction of strain in the doping and alloying process. It is very important to introduce strain regulation into 2D semiconductor materials to broaden the application field of 2D materials. For example, as shown in Figure 8.1, graphene and monolayer MoS_2 can withstand strain up to 25%.[22–24] Mohr et al. found that strain causes splitting in the Raman spectrum of graphene, and the degree of splitting depends on the strength and direction of strain.[25] Scalise et al. reported that strain affects the electronic structure and vibration characteristics of a honeycomb MoS_2.[26] Johari's group reported the effect of tensile or compressive strain on the electronic properties of transition metal sulfides. The study showed that the band gap changes significantly under strain and showed the transition from indirect band gap to direct band gap and then to indirect band gap.[27, 28] Moreover, the carrier mobility of phosphene can also be improved by applying compression strain perpendicular to plane.[29, 30] Due to its flexibility and robustness, strain engineering can also achieve effective control of heat transport properties.[31]

Electric field: applying external electric field is a common means to control the properties of materials by changing the external environment. The introduction of external electric field can redistribute the electrons of two-dimensional materials, so as to change the energy band structure and effectively adjust various properties. The research on graphene, black phosphorus, and some related heterojunctions proves the correctness of applying electric field to regulate the properties of materials.[32, 33] Additionally, Li et al. studied the changes in the properties of MoS_2/arsenene under the external electric field, and found that the band gap value of the heterojunction varies linearly with the intensity of the electric field under the action of the electric field. At the same time, the band gap type changes from Type-II to Type-I. When the electric field intensity is large enough, the semiconductor metal transformation will also occur.[34] And the electronic structure and optical properties of two dimensional transition metal chalcogenides are also influenced by electric field.[35]

Magnetic field: magnetic field generates cyclotron resonance and Shubnikov de Haas effect through the action on charge carriers in matter and the response of electron Landau energy, and magnetization, magnetic phase transition, magnetic resonance, and other effects due to its action on the magnetic moment formed by electron orbit and spin. Therefore, as one of the important thermodynamic parameters as well as temperature and pressure, magnetic field can change the internal energy of a material and is an effective condition for studying the physical phenomenon and mechanism of materials. Application of magnetic field is the most tunable approach to controlling valley pseudospins.[36] The valley Zeeman effect has been corroborated in the polarization-resolved photoluminescence spectrum of monolayer MoS_2, $MoSe_2$, WSe_2 (shown in Figure 8.2) , WS_2, and $MoTe_2$ in external magnetic field.[37–44]

8.3   2D XENES MATERIALS UNDER STRAIN

Generally, two-dimensional materials are able to withstand greater strains than three-dimensional bulk materials. Both Pang's and Wang's results demonstrate that the mechanical properties of borophene along armchair and zigzag direction are highly anisotropic under strain regulation. Theoretically speaking, the strain along the y direction will change the bond length and thus affect the delocalization, while the strain along the x direction will change the buckling and the system becomes more planar. Moreover, the phonon spectrum is dynamically unstable regardless of the biaxial strain applied along armchair or zigzag direction, while

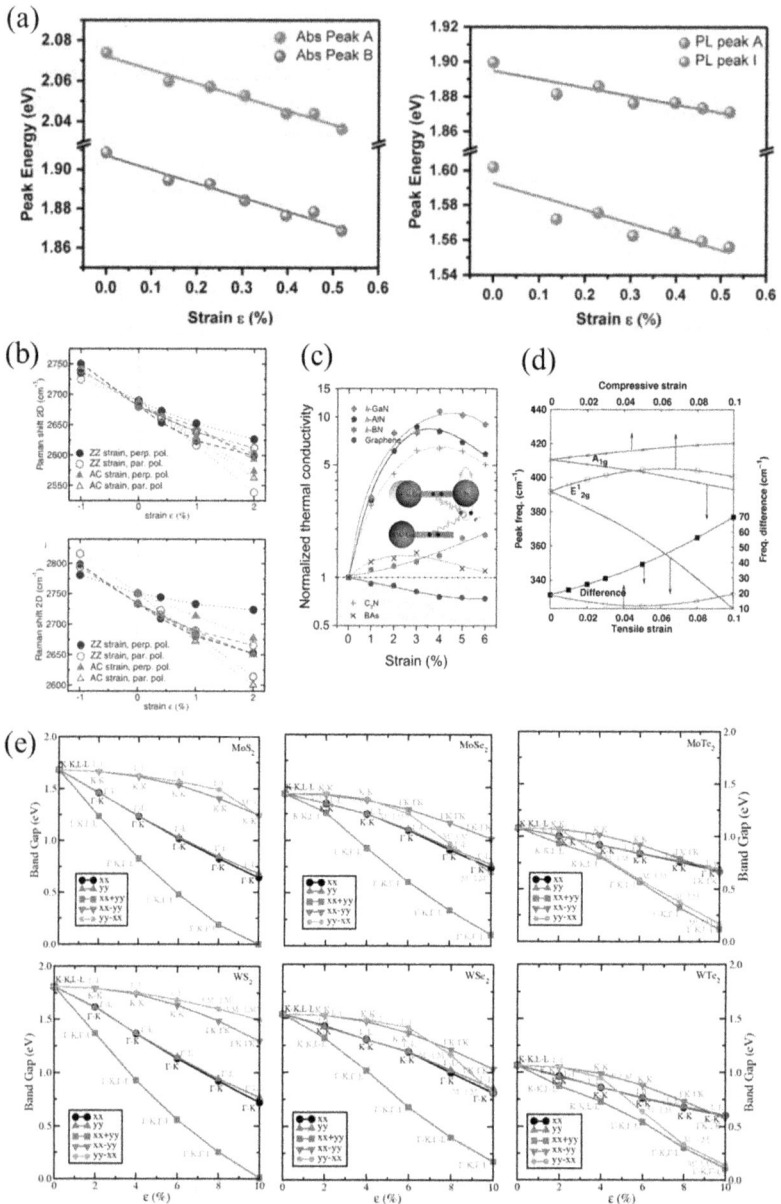

FIGURE 8.1 (a) Strain dependence of the A and B peak energies determined from absorption and of the A and I peak energies determined from PL are summarized. Reproduced with permission.[24] Copyright 2013, American Chemistry Society. (b) Mapping of the Raman 2D contributions for inner and outer processes for the laser energies 1.5 eV(up) and 2.4 eV(down). Circles triangles correspond to strain in zigzag armchair directions. Open closed symbols denote parallel perpendicular polarization with respect to the strain

the phonon spectrum is stable under uniaxial strain applied along zigzag direction.[45] Regarding borophene as a pristine 2D monolayer superconductor, applying tensile strain can increase the superconducting transition temperature (T_c) to 27.4 K.[46] Li et al. investigated the effects of strain on the structure stability, electron-phonon coupling (EPC), and superconductivity of two types of monolayer borophenes, and it was shown that the electron–phonon coupling (EPC) in the buckled triangle borophene can be significantly enhanced by the suitable strain (−2–3%). With the increase of tensile strain, the superconducting transition temperature changes in a U-shaped curve. As shown in Figure 8.3 (a–b), although borophene itself is anisotropic, the influence of uniaxial strain on its superconducting characteristics is isotropic (Figure 8.3 (a–b)).[47] Mahdi et al. studied the effects of strain on 2D borophene with β_{12}, χ_3 and striped. They found that the electronic structures and elastic constants of these borophene showed nonlinear changes when tensile strains and compressive strains were applied in different directions, which indicated that their electronic behavior was anisotropic.[48] When the tensile strain is greater than 3% and the compressive strain is greater than 5%, the phonon spectra of borophene shows negative modes, indicating that their structures are unstable under these strain conditions.[49] It shows that with net charge doping and in-plane tensile strain, the borophene becomes thermodynamic stable in ideal plat nature, because the bonding characteristic is modified.[50] Wang et al. showed that applying tensile strain in the b direction could increase the buckling height of borophene, resulting in an out-of-plane negative Poisson's ratio that makes the boron sheet exhibit better mechanical elasticity in b direction (Figure 8.3 (b)).[51] Meanwhile, reports have also demonstrated that strain can regulate the polarization direction of incident light, thus affecting the response of borophene to light.

Pholene is an excellent 2D material with high carrier mobility and adjustable band gap. The research of Drissi et al. found that uniaxial strains modulate the band gap, the buckling parameter, and local charge in elastic and metastable regions. In the elastic range, armchair strain flattens more phosphorene compared to zigzag strain. And the variation of band gap shows semiconductor-to-metal transition at 29% and 26% along x- and y-direction respectively (Figure 8.3 (c)).[52] For bilayer phosphoene, the direct bandgap is transformed into an indirect bandgap by applying compressive strain. When the strain increases to 7.47%, the bilayer phosphoene is transformed from an indirect bandgap semiconductor to a metal. Additionally, by calculating the carrier effective mass and mobility under

---

FIGURE 8.1 (CONTINUED)

direction. Reproduced with permission.[25] Copyright 2010, American Physical Society. (c) The anomalous strain enlarged κ of h-(B/Al/Ga)N, h-BAs, and C_3N, in sharp contrast to grapheme. All the κ are normalized to their respective intrinsic κ without strain. (Inset) Schematic of the strain weakened interactions of lone-pair N-s electrons with the bonding electrons. Reproduced with permission.[31] Copyright 2018, Elsevier. (d) Frequency of the Raman modes and their difference versus applied strain. Reproduced with permission.[26] Copyright 2014, Elsevier. (e) Band gap of monolayer TMDs with respect to strain, ε, which varies from 0 to 10%.[27] Copyright 2012, American Chemistry Society.

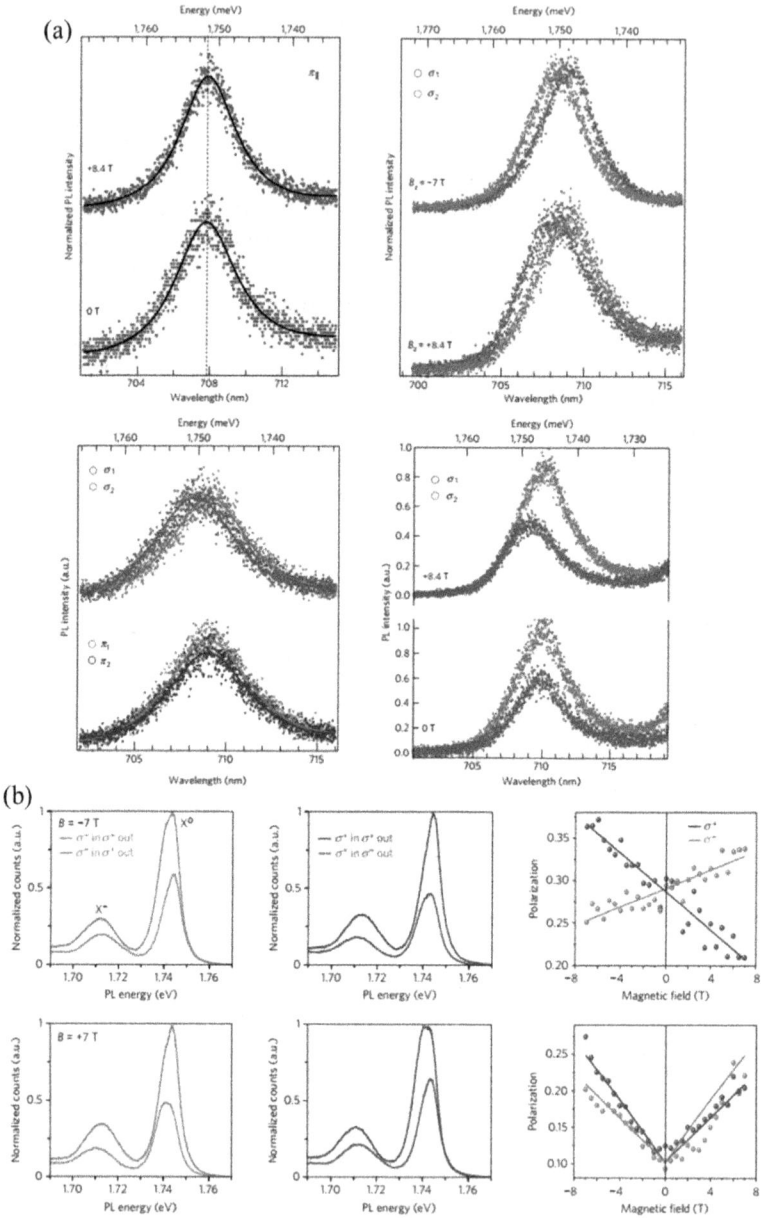

FIGURE 8.2 The extracted valley Zeeman splitting versus magnetic field of WSe$_2$. Reproduced with permission.[37, 39] Copyright 2015, Springer.

strain control, it is found that the transport properties do not change significantly as a function of strain.[53] The uniaxial strain and biaxial strain have different effects on the black phosphoene, as shown in (Figure 8.3 (d)). Uniaxial strain and biaxial strain have different effects on black phosphoene. Under the uniaxial strain of 0–8% along y direction, the band gap of black phosphoene changes from direct to indirect, while under the biaxial strain of 0–8%, the band gap of black phosphoene maintains direct.[54, 55] Moreover, the strain engineering improves the photocatalysis properties by adjusting the band gap of phosphorene into the visible light wavelength range, which will facilitate the absorption of the visible light and thus increase the efficiency of the photocatalytic water splitting.[56] Because the electron–phonon coupling is more significantly enhanced by the biaxial strain than the uniaxial strains, when the biaxial strain reaches 4.0%, the superconducting transition temperature T_c increases sharply from 3 K to 16 K.[57]

In silicene, the band gap opens for smaller uni- and biaxial strains and increases of strain beyond 8% silicene changed into metal. It is worth mentioning that the imaginary part of the dielectric function shows that the inter-band transitions are red-shifted for uni- and biaxial tensile strains and are blue shifted for uni- and biaxial compressive strains.[58] Qin et al. reported that silicene undergoes the semimetal to metal transition at the strain of 7%, and the Fermi velocity changes very little before this critical strain. The work function is found to first increase with the increasing strain and then nearly saturate.[59] Theoretically, strain engineering can effectively regulate the electronic and optical properties of materials, while thermoelectric properties are generally considered to be the least sensitive to mechanical strains. Xie et al.[60] found that the thermal conductivity of silicene can increase dramatically within 10% tensile strain, mainly because silicene has a buckling structure, and the structure tended to be planarized due to strain engineering, which resulted in the increase of the acoustic phonon lifetime (Figure 8.3 (e–h)). Wang et al.[61] found that due to the valley-opposite gauge field induced by the strain, the strained silicene with a superlattice structure exhibits an angle-resolved valley and spin filtering effect when the spin–orbit interaction is considered. Moreover, compression strain applied to bilayer silicene and germanene can transform conventional metals into topological insulators. In semimetallic germanene, the Dirac point was predicted to shift above the Fermi level with tensile strain of up to 16%, inducing p-type behavior due to decreased sp hybridization and increased Ge–Ge bond length.[62, 63]

Different from silicene and germanene, uniaxial strain can regulate SOC-induced band gap in stanene without changing the quantum spin Hall (QSH) state.[64, 65] Broek et al. suggested that applying lateral strain on stanene under out-of-plane electric field can open up the bandgap up to 0.21 eV, while Modarresi et al. suggested that the SOC bandgap of ≈0.07 eV in stanene closes under applied strain. And Fu et al. studied stanene and predicted a metal–insulator transition under critical biaxial strain.[66–68] In particular, planar indiene always keeps its metal form under the biaxial strain of tension or compression, while buckled indiene changes from an indirect semiconductor to a metal under the strain. Theoretical calculations from Xu et al.[69] demonstrated that monolayer arsenene was transformed from indirect to

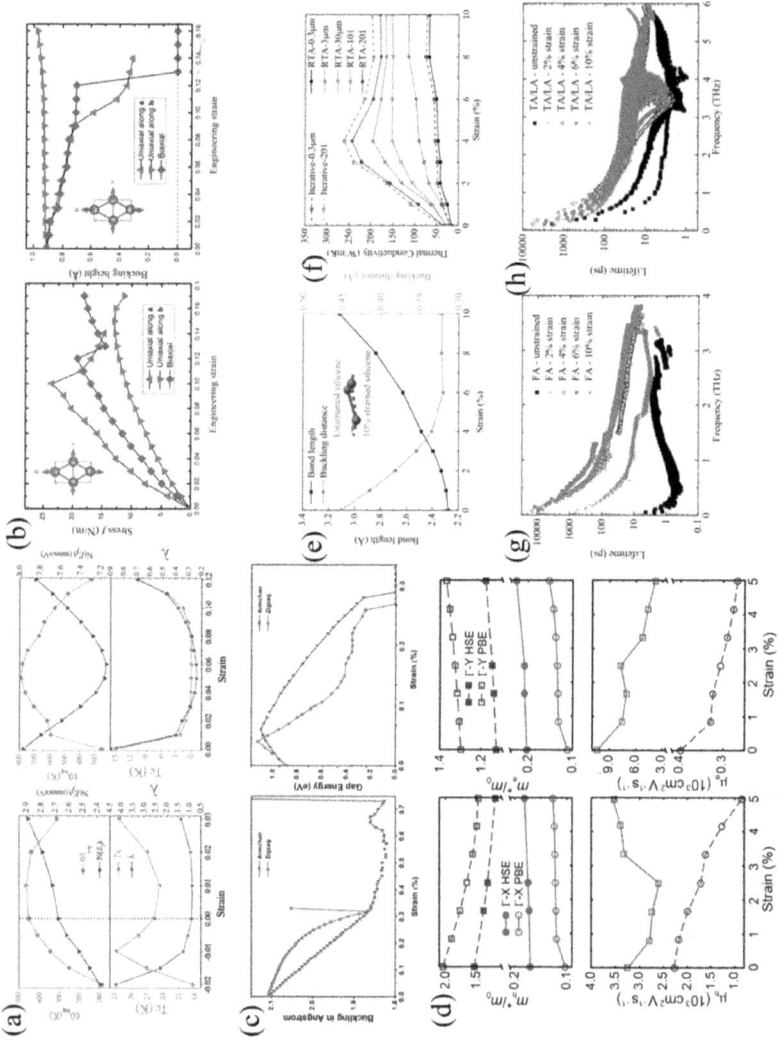

FIGURE 8.3

direct band gap semiconductor by inducing uniaxial tensile strain along armchair and zigzag directions. Moreover, they also pointed out that with the increase of strain, the light absorption characteristics of arsenene showed obvious red shift. Strain engineering can adjust the bandgap of either puckered or buckled arsenene. Applying a strain of only 1% is sufficient to convert the puckered arsenene into a direct bandgap semiconductor, while the system goes from semiconductor to metallic for the values of strains beyond –10% and 12%.[70] Remarkably, strain modulates not only the band gap of monolayer arsenene, but also the effective mass and the emergence of Dirac-like cone with in-plane strain.[71] For group-V elements, monolayer antimonene was predicted to have a 2.3 eV indirect bandgap that undergoes an indirect-to-direct transition under strain.[72] It is reported that β-antimonene could sustain large (tensile) strain (up to 20%), and the band inversion was demonstrated to take place in the vicinity of the Γ point as the strain grows larger than a value that ranges from ≈7% to ≈18%.[73] In addition, the strain engineering also affects the magnetic properties of monolayer antimonene with vacancy doping. When the tensile strain is above 4%, the total magnetic moment of the vacancy-doped model is unchanged, but is transformed into an antiferromagnetic semiconductor. The magnetic moment is enhanced to 3 μ_B and the metastable ferromagnetic semiconductor is formed with tensile strain exceeding 7% (Figure 8.4 (a–b)).[74] Zhao et al.[75] proposed that the strain-driven band inversion in the vicinity of the Fermi level is responsible for the quantum phase transition in antimoene.

The monolayer tellurene can be transformed from centrosymmetric β-phase to non-centrosymmetric α-phase under strain, and the phase transition point is when the tensile strain is 0.5%. Meanwhile, the electrical conductance of β-tellurene changes from armchair to zigzag, in the strain range of –1% to 0%. In particular, the compressive strain increases the electron mobility along the armchair direction, while the tensile strain increases the electron mobility even more along the zigzag direction, and the anisotropic electron effective mass tensor and the corresponding mobility can be rotated by 90° (Figure 8.4 (c)).[76–78] Zhu et al.[79] found

---

FIGURE 8.3 (OPPOSITE) (a–b) Superconducting transition temperature T_c, EPC constant λ, DOS of the FS $N(E_F)$, and logarithmically averaged phonon frequency ω_{log} versus the strain ε (a) and uniaxial strains (b) for the (left) buckled triangle borophene and (right) β_{12} borophene, respectively. The curves marked with the hollow and solid symbols represent the uniaxial strain applied along the a and b directions in (b), respectively. Reproduced with permission.[47] Copyright 2018, American Chemical Society. (c) Strain variations of buckling parameter and band gap energy along x and y directions for pure phosphorene. Reproduced with permission.[52] Copyright 2017, Elsevier. (d) The band gap of 2D phosphorene as a function of strain applied in the zigzag (left) and armchair (right) directions, respectively. (e) Bond length and buckling distance as a function of strain. Inset: Primitive unit cell structures for unstrained and 10% strained silicene. (f) Thermal conductivity of infinite and finite-size silicene as a function of strain computed. (g–h) Lifetime of FA phonons (g) and LA/TA phonons (h) as a function of frequency for 0%, 2%, 4%, 6%, and 10% strain. Reproduced with permission.[53, 60] Copyright 2016, American Physical Society.

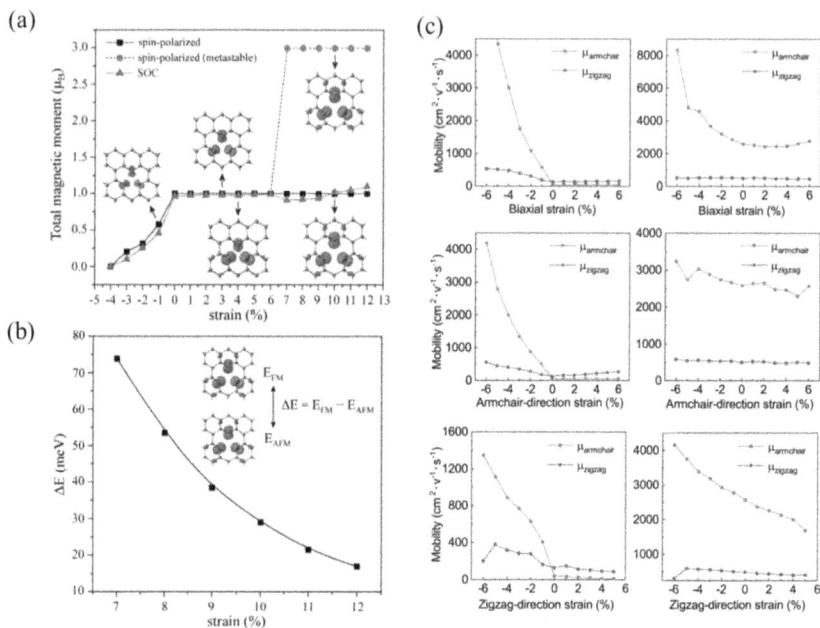

FIGURE 8.4 (a) Magnetic property (inserts are spin density distribution around the vacancy at –1%, 3%, 4%, and 10% strained states with isosurface = ± 0.01 e/Å³. (b) Energy difference between stable AFM and metastable FM states. Reproduced with permission.[74] Copyright 2019, Elsevier. (c) Electron mobilities of β-Te along armchair and zigzag directions for strain. Reproduced with permission.[77] Copyright 2019, Royal Society of Chemistry.

that CBM shifts down gradually towards the Fermi Level as the tensile strain increases from 0% to 6%, whereas the energy of VBM almost does not change, resulting in the reduced band gap, and the band gap reduced to 0.86 eV under the tensile strain of 6%.

The corresponding indiene monolayers are in three different allotropic forms: planar, puckered, and buckled, it is reported that the planar allotrope remains metallic under both compressive and tensile strain, while buckled allotrope changes from an indirect semiconductor to a metal.[80] For the indiene, phonon frequencies remain real throughout the Brillouin zone for the entire range of strains in the armchair direction under tensile strain, while in the zigzag direction the frequencies remain real up to 14% strain. Analogously, it is shows that aluminene unstable phonon mode becomes stable only for strains of 14 and 18% along armchair direction. However, when strained in zigzag direction, the phonon mode is stable as soon as 4% strain is applied.[81] Yeoh et al. found that aluminene remains stable even when tensile strain reaches 7%. On the other hand, the stability of the structure cannot be guaranteed under compressive strain.[82] Strain engineering can not only change the crystal structure and electronic structure of 2D Xenes but also further regulate its magnetic characteristics, optical properties, thermoelectric effects, and transport properties.

8.4   2D XENES MATERIALS UNDER ELECTRIC FIELD

From the perspective of regulating the electronic properties of materials, the application of electric field is one of the important means to open the band gap and modulate the properties. In theoretical predictions, unlike graphene which is unaffected by external vertical electric field, the band structure of both silicene and germanene can be regulated by applying an out-of-plane vertical electric field (Figure 8.5 (a)).[83, 84] Also by increasing the electric field magnitude the band gap increases in the silicene. In monolayer silicene, the bandgap varies linearly with electric field intensity, and the total band gap of up to 18 meV was predicted. In Jose and Datta's research,[85–87] the structural properties of silicene and graphene were compared. In fact, it is found that the opening of band gap in silicene is the effect of external vertical electric field. And the essential effect of applying an external electric field is to break the symmetry between the A and B sublattices of the silicene honeycomb structure, thus opening the band gap [88] under the electric field and causing other interesting phenomena.[89, 90]

In addition, studies have shown that there are single layer and few layers in 2D crystals, and the local electric field can effectively change the carrier sign. Ni et al.[91] studied the field-effect transistor based on silicene, and achieved a low switching ratio of 4.2 by applying a vertical electric field of 1V/ Å at the condition of 300 K. This also allows the electronic properties of silicene to be tweaked for use in field effect transistors. Additionally, an external electric field can also reduce the lattice thermal conductivity of 2D silicene. Furthermore, due to the valley-opposite gauge field induced by the strain, the strained silicene with a superlattice structure exhibits an angle-resolved valley and spin filtering effect when the spin–orbit interaction is considered.[92] According to the work of Qin et al.,[93] the main principle is that the charge density is redistributed and the screened potential is generated with an external electric field, which leads to the renormalization of the interaction between silicene atoms and further generates electron-phonon coupling to modulate the nonharmonic motion of phonons (Figure 8.5 (e)). Shu et al.[94] systematically investigated the effect of electric field on the photoelectric properties of hexagonal arsenene (β-As) and antimonene (β-Sb). Appling the electric field of 0.51 V/Å and 0.50 V/Å to β-As and β-Sb, respectively, resulted in the transition from indirect-to-direct band gap. In addition, the high electric field of 0.6V/ Å can enhance the photon absorption of β-As and β-Sb between 1.0 and 2.5 eV, indicating that the external electric field can regulate the photoexcitation of β-As and β-Sb. (Figure 8.5 (b–c))

For tetragonal germanene, the applied electric field always increases the degeneracy of Dirac points A and B, resulting in a finite band gap (Figure 8.5 (d)). It also shows a slight self-doping effect under the electric field of 0.125 and 0.5 V/Å.[95] It is also predicted that the vertical electric field can linearly increase the band gap value and transform semimetallic silicene, germanene, and stanene monolayers

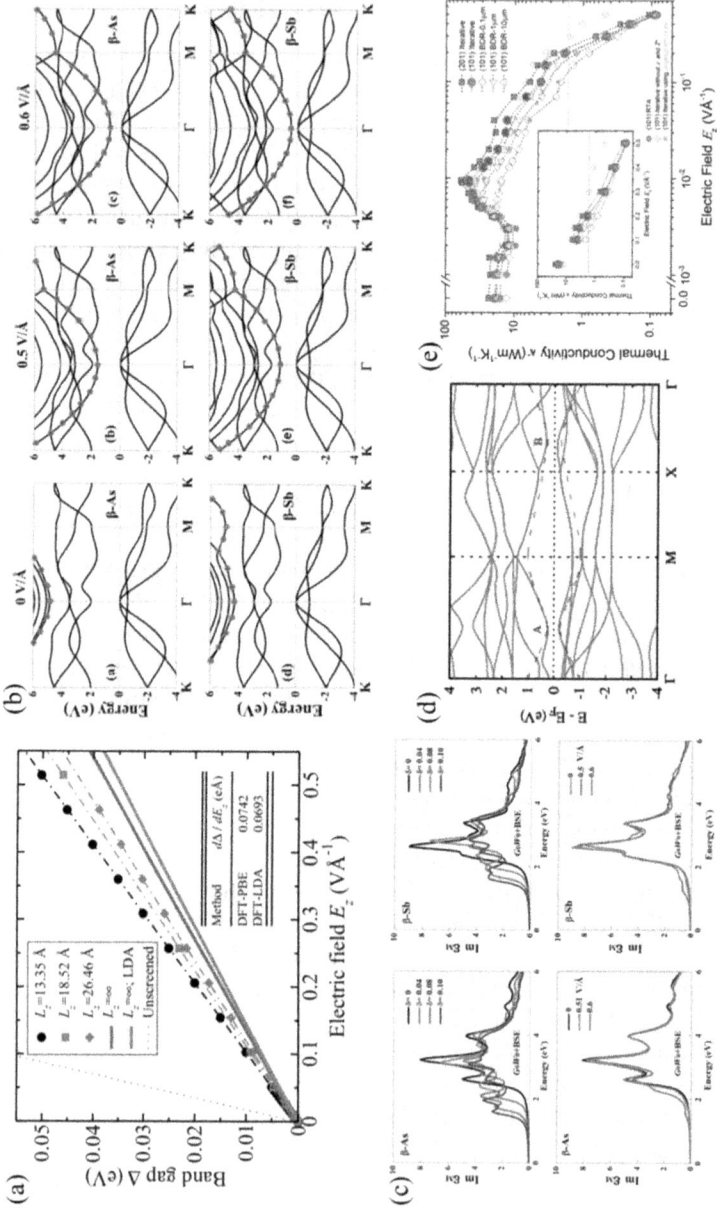

FIGURE 8.5

into semiconductors. Fu et al.[68] studied the electronic and topological properties of stanene and predicted a topological nontrivial–trivial transition under a critical vertical electric field. Liu et al.[96] demonstrated that the electric field can effectively promote the band spin-degeneracy splitting at the monolayer bismuthene valence band maximum, and the band gap decreases with the increase of the electric field. It is reported by Ghosh et al. that the position and carrier velocities of the spin-polarized Dirac cones of multilayered phosphorene could be controlled by the strength of the external transverse electric field.[97] The application of a vertical electric field onto blue phosphorene even can induce a semiconductor–metal transition when the electric field is as large as about 0.7 V/Å. The application of an electric field effectively modulates the material structure and properties, and also broadens the application design of 2D Xenes.

8.5 2D XENES MATERIALS UNDER MAGNETIC FIELD

The application of external magnetic field can regulate the electric polarization, optical polarization, temperature, geometric shape, and other macroscopic properties of magnetic materials to achieve magneto-electric, magneto-optical, magneto-thermal, magneto-elastic, and other effects. These effects are the physical basis of magnetic function devices such as magnetic detectors, magneto-optical Kerr apparatus, magnetic refrigerators, etc. Considering that the macroscopic physical properties of materials are closely related to the microscopic electronic structure, the most intuitive idea is to directly regulate the electronic energy band structure through the magnetic field, so as to change the electrical and optical properties of materials. In addition, the photoconductivity, heat transport, and other properties of 2D Xenes materials can be regulated by applying an external magnetic field.

Yu et al.[98] (Figure 8.6 (a–b)) proposed a Hamiltonian model for the low energy band structure of monolayer arsenene and studied the effect of external magnetic fields on monolayer arsenene using the tight-bound model and first-principles methods. It was found that with the increase of magnetic field intensity,

---

FIGURE 8.5 (CONTINUED) (a) DFT gap against applied electric field Ez for silicene with a plane-wave cutoff energy of 816 eV and a 53 × 53 **k**-point grid. Unless otherwise stated, the PBE functional was used. The box length in the z direction was varied from Lz = 13.35Å to 26.46Å. Reproduced with permission.[83] Copyright 2012, American Physical Society. (b) Band structures at the external electric field of 0, 0.5, and 0.6 V/Å. (c) Optical absorption spectra of β-As and β-Sb for the light polarized along a direction under biaxial tensile strain and externally electric field. Reproduced with permission.[94] Copyright 2018, Royal Society of Chemistry. (d) Band structure of tetragonal germanene with vertical electric field of 0.75 V/Å. Solid and dashed lines indicate the band spectra corresponding to DFT and TB techniques. Reproduced with permission.[95] Copyright 2020, Royal Society of Chemistry. (e) Lattice thermal conductivity (κ) of silicene at 300 K as a function of the strength of the external electric field (Ez). (Inset) The κ versus several large Ez values plotted on a linear scale. Reproduced with permission.[93] Copyright 2017, Royal Society of Chemistry.

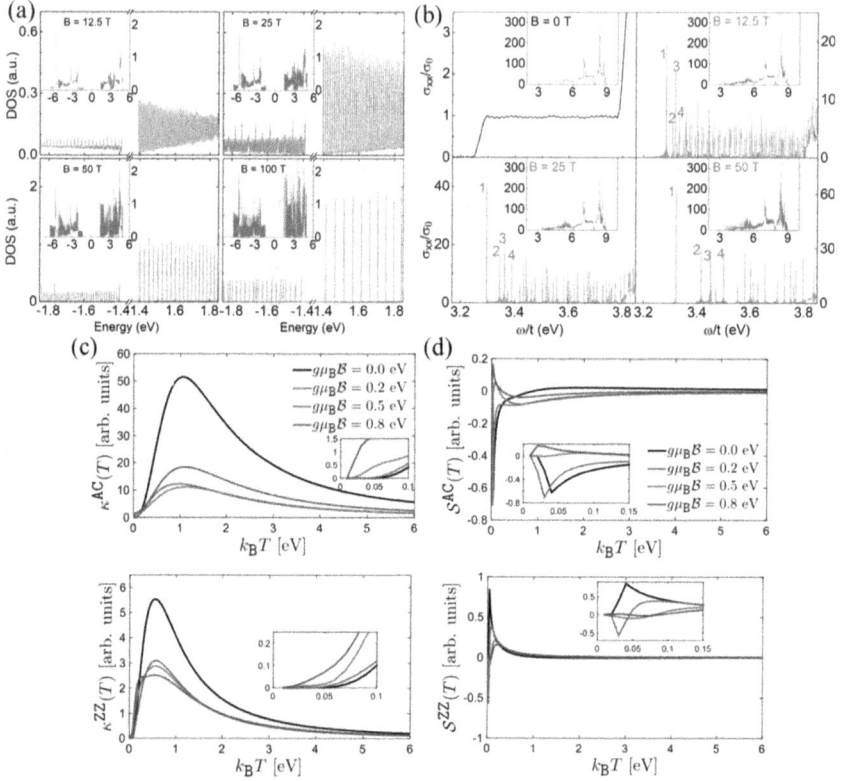

FIGURE 8.6 (a) DOS of the LLs of electrons and holes of ML-As under perpendicular magnetic field. (b) Magneto-optical properties of ML-As. The optical conductivity spectrum of ML-As is calculated with B = 0T, 12.5T, 25T, and 50T, respectively. The inset shows the optical conductivity in a much wider energy window. $\sigma 0 = e^2/4\hbar$ is the universal optical conductivity. The first four peaks are present in orange for reference. Reproduced with permission.[98] Copyright 2018, American Physical Society. (c) The electronic contribution of thermal conductivity along the AC and ZZ direction in the presence of applied Zeeman magnetic field. (d) ψ^{AC} (T) and ψ^{ZZ} (T) at various magnetic fields in monolayer black phosphorene. Reproduced with permission.[99] Copyright 2019, Elsevier.

the electrical conductivity spectrum of monolayer arsenene showed a blue shift, which proved that the electronic and magneto-optical properties of monolayer arsenene could be adjusted by the vertical magnetic field. Thermoelectric (TE) materials have attracted attention in recent years due to their ability to convert thermal energy directly to electrical energy or electric energy directly to thermal energy. The electrical and thermal properties of phosphoenes in 2D Xenes have been studied. It has been shown that in the presence of magnetic field, using Green's function method and tight bound model, the change rules of the magnetic field-dependent electrical conductivity of phosphoene in the armchair and zigzag direction are different; (Figure 8.6 (c–d)) further explains the regulation of magnetic field on the magnetic field-induced conductivity of phosphoene.[99] Romera et al.[100] studied the flutter effect in a single layer of silicon under a vertical magnetic field, by considering wave packets with a Gaussian population of both positive and negative energy levels. The introduction of external magnetic fields mainly provides a new design solution for 2D Xenes in the magnetic field.

8.6 SUMMARY

In summary, the 2D Xenes have a number of interesting and outstanding properties that make them useful in a variety of fields. At the same time, in order to be better applied in specific fields, external fields are introduced to break through the limitations of materials in structure, stability, electricity, magnetism, mechanics, and optics, etc. In this paper, the characteristics, significance, and application effects of common outfields are summarized, and the influences of different outfields on two-dimensional Xenes are also summarized. The results show that strain, electric field, and magnetic field can exert long-term mechanisms on the electronic structure, electrical properties, magnetic properties, spin-orbit coupling, quantum Hall effect, and other aspects of two-dimensional Xenes, which further expands the application of this kind of material in different fields.

REFERENCES

1. Novoselov, K. S. et al., Two-dimensional gas of massless Dirac fermions in graphene. *Nature*, 438, 7065, 2005.
2. Zhang, Y., Tan, Y. W., Stormer, H. L. & Kim, P., Experimental observation of the quantum Hall effect and Berry's phase in graphene. *Nature*, 438, 7065, 2005.
3. Stankovich, S. et al., Graphene-based composite materials. *Nature*, 442, 7100, 2006.
4. Michel, K. H. & Verberck, B., Theory of elastic and piezoelectric effects in two-dimensional hexagonal boron nitride. *Physical Review B*, 80, 22, 2009.
5. Wang, Q. H., Kalantar-Zadeh, K., Kis, A., Coleman, J. N. & Strano, M. S., Electronics and optoelectronics of two-dimensional transition metal dichalcogenides. *Nature Nanotechnology*, 7, 11, 2012.
6. Wang, Z. T., Chen, Y., Zhao, C. J., Zhang, H. & Wen, S. C., Switchable dual-wavelength synchronously Q-Switched Erbium-Doped fiber laser based on graphene saturable absorber. *IEEE Photonics Journal*, 4, 3, 2012.
7. Huang, Z. et al., Wall-like hierarchical metal oxide nanosheet arrays grown on carbon cloth for excellent supercapacitor electrodes. *Nanoscale*, 8, 27, 2016.
8. Li, Z. et al., High-performance photo-electrochemical photodetector based on liquid-exfoliated few-layered InSe nanosheets with enhanced stability. *Advanced Functional Materials*, 28, 16, 2018.
9. Zhuang, J. et al., Band gap modulated by electronic superlattice in Blue Phosphorene. *ACS Nano*, 12, 5, 2018.
10. Ezawa, M., Monolayer topological insulators: Silicene, Germanene, and Stanene. *Journal of the Physical Society of Japan*, 84, 12, 2015.
11. Tao, W. et al., Emerging two-dimensional monoelemental materials (Xenes) for biomedical applications. *Chemical Society Reviews*, 48, 11, 2019.

12. Guo, Q. et al., Black Phosphorus mid-infrared photodetectors with high gain. *Nano Letters*, 16, 7, 2016.
13. Kong, X., Liu, Q., Zhang, C., Peng, Z. & Chen, Q., Elemental two-dimensional nanosheets beyond graphene. *Chemical Society Reviews*, 46, 8, 2017.
14. Kripalani, D. R., Kistanov, A. A., Cai, Y., Xue, M. & Zhou, K., Strain engineering of antimonene by a first-principles study: Mechanical and electronic properties. *Physical Review B*, 98, 8, 2018.
15. Qiu, M. et al., Omnipotent phosphorene: A next-generation, two-dimensional nano-platform for multidisciplinary biomedical applications. *Chemical Society Reviews*, 47, 15, 2018.
16. Zhang, S. et al., Recent progress in 2D group-VA semiconductors: from theory to experiment. *Chemical Society Reviews*, 47, 3, 2018.
17. Li, W. et al., Molybdenum diselenide – black phosphorus heterostructures for electrocatalytic hydrogen evolution. *Applied Surface Science*, 467–468, 2019.
18. Tian, X. et al., Modulating Blue Phosphorene by synergetic codoping: Indirect to direct gap transition and strong bandgap bowing. *Advanced Functional Materials*, 29, 11, 2019.
19. Glavin, N. R. et al., Emerging applications of elemental 2D materials. *Advanced Materials*, 32, 7, 2020.
20. Wang, T. et al., Xenes as an emerging 2D monoelemental family: Fundamental Electrochemistry and Energy Applications. *Advanced Functional Materials*, 30, 36, 2020.
21. Qiao, H. et al., Tunable electronic and optical properties of 2D monoelemental materials beyond graphene for promising applications. *Energy & Environmental Materials*, 4, 4, 2021.
22. Fang Ying, S. N.-H. & Dlott, D. D., Measurement of the distribution of site enhancements in surface-enhanced Raman Scattering. *Science*, 321, 2008.
23. Bertolazzi, S., Brivio, J. & Kis, A., Stretching and breaking of ultrathin MoS2. *ACS Nano*, 5, 12, 2011.
24. He, K., Poole, C., Mak, K. F. & Shan, J., Experimental demonstration of continuous electronic structure tuning via strain in atomically thin MoS2. *Nano Letters*, 13, 6, 2013.
25. Mohr, M., Maultzsch, J. & Thomsen, C., Splitting of the Raman2Dband of graphene subjected to strain. *Physical Review B*, 82, 20, 2010.
26. Scalise, E., Houssa, M., Pourtois, G., Afanas'ev, V. V. & Stesmans, A., First-principles study of strained 2D MoS2. *Physica E: Low-dimensional Systems and Nanostructures*, 56, 2014.
27. Johari, Priya, & Shenoy, Vivek B., Tuning the electronic properties of semiconducting transition metal Dichalcogenides by applying mechanical strains. *ACS Nano*, 6, 2012.
28. Zhang, Z., Wang, J., Song, C., Mao, H. & Zhao, Q., Tuning band gaps of transition metal Dichalcogenides WX2 (X = S, Se) Nanoribbons by external strain. *Journal of Nanoscience and Nanotechnology*, 16, 8, 2016.
29. Morgan Stewart, H., Shevlin, S. A., Catlow, C. R. & Guo, Z. X., Compressive straining of bilayer phosphorene leads to extraordinary electron mobility at a new conduction band edge. *Nano Letters*, 15, 3, 2015.
30. Fang, R. et al., Strain-Engineered ultrahigh mobility in Phosphorene for Terahertz Transistors. *Advanced Electronic Materials*, 5, 3, 2019.
31. Qin, G., Qin, Z., Wang, H. & Hu, M., Lone-pair electrons induced anomalous enhancement of thermal transport in strained planar two-dimensional materials. *Nano Energy*, 50, 2018.

32. Yan, J., Zhang, Y., Kim, P. & Pinczuk, A., Electric field effect tuning of electron-phonon coupling in graphene. *Physical Review Letters*, 98, 16, 2007.
33. Qiao, J., Kong, X., Hu, Z.X., Yang, F. & Ji, W., High-mobility transport anisotropy and linear dichroism in few-layer black phosphorus. *Nature Communications*, 5, 2014.
34. Li, W., Wang, T., Dai, X., Ma, Y. & Tang, Y., Effects of electric field on the electronic structures of MoS2/arsenene van der Waals heterostructure. *Journal of Alloys and Compounds*, 705, 2017.
35. Liu, F., Zhou, J., Zhu, C. & Liu, Z., Electric field effect in two-dimensional transition metal Dichalcogenides. *Advanced Functional Materials*, 27, 19, 2017.
36. Liu, W., Luo, C., Tang, X., Peng, X. & Zhong, J., Valleytronic properties of monolayer WSe2 in external magnetic field. *AIP Advances*, 9, 4, 2019.
37. Aivazian, G. et al., Magnetic control of valley pseudospin in monolayer WSe2. *Nature Physics*, 11, 2, 2015.
38. MacNeill, D. et al., Breaking of valley degeneracy by magnetic field in monolayer MoSe2. Physical Review Letters, 114, 3, 2015.
39. Srivastava, A. et al., Valley Zeeman effect in elementary optical excitations of monolayer WSe2. *Nature Physics*, 11, 2, 2015.
40. Arora, A. et al., Valley Zeeman splitting and valley polarization of neutral and charged excitons in monolayer MoTe2 at high magnetic fields. *Nano Letters*, 16, 6, 2016.
41. Plechinger, G. et al., Excitonic valley effects in monolayer WS2 under high magnetic fields. *Nano Letters*, 16, 12, 2016.
42. Stier, A. V., McCreary, K. M., Jonker, B. T., Kono, J. & Crooker, S. A., Exciton diamagnetic shifts and valley Zeeman effects in monolayer WS_2 and MoS_2 to 65 Tesla. *Nature Communications*, 7, 2016.
43. Koperski, M. et al., Orbital, spin and valley contributions to Zeeman splitting of excitonic resonances in $MoSe_2$, WSe_2 and WS_2 Monolayers. *2D Materials*, 6, 1, 2018.
44. Wu, Y. J. et al., Valley Zeeman splitting of monolayer MoS2 probed by low-field magnetic circular dichroism spectroscopy at room temperature. *Applied Physics Letters*, 112, 15, 2018.
45. Li, D., Chen, Y., He, J., Tang, Q., Zhong, C. & Ding, G., A review of thermal transport and electronic properties of borophene. *Chinese Physics B*, 27, 2008.
46. Xiao, R. C., Shao, D. F., Lu, W. J., Lv, H. Y., Li, J. Y. & Sun, Y. P., Enhanced superconductivity by strain and carrier-doping in borophene: A first principles prediction. *Applied Physics Letters*, 109, 2016.
47. Li, G., Zhao, Y., Zeng, S., Zulfiqar, M. & Ni, J., Strain effect on the superconductivity in Borophenes. *The Journal of Physical Chemistry C*, 122, 29, 2018.
48. Faghihnasiri, M. et al., Nonlinear elastic behavior and anisotropic electronic properties of two-dimensional borophene. *Journal of Applied Physics*, 125, 14, 2019.
49. Bhattacharyya, G., Mahata, A., Choudhuri, I. & Pathak, B., Semiconducting phase in borophene: role of defect and strain. *Journal of Physics D: Applied Physics*, 50, 40, 2017.
50. Yin, Y. et al., Phonon stability and phonon transport of graphene-like borophene. *Nanotechnology*, 31, 31, 2020.
51. Wang, H. et al., Strain effects on borophene: Ideal strength, negative Possion's ratio and phonon instability. *New Journal of Physics*, 18, 7, 2016.
52. Drissi, L. B., Sadki, S. & Sadki, K., Phosphorene under strain: Electronic, mechanical and piezoelectric responses. *Journal of Physics and Chemistry of Solids*, 112, 2018.

53. Manjanath, A., Samanta, A., Pandey, T., Singh, A. K., Semiconductor to metal transition in bilayer phosphorene under normal compressive strain. *Nanotechnology*, 26, 2015.

54. Zhang, R.-Y., Zheng, J.-M. & Jiang, Z.-Y., Strain effects on properties of Phosphorene and Phosphorene Nanoribbons: A DFT and tight binding study. *Chinese Physics Letters*, 35, 1, 2018.

55. Peng, X., Qun, W. & Andrew, C. Strain-engineered direct-indirect band gap transition and its mechanism in two-dimensional phosphorene. *Physical Review B*, 90, 8, 2014.

56. Sa, B., Li, Y.-L., Qi, J., Ahuja, R. & Sun, Z., Strain engineering for Phosphorene: The potential application as a photocatalyst. *The Journal of Physical Chemistry C*, 118, 46, 2014.

57. Ge, Y., Wan, W., Yang, F. & Yao, Y., The strain effect on superconductivity in phosphorene: a first-principles prediction. *New Journal of Physics*, 17, 3, 2015.

58. Mohan, B., Kumar, A. & Ahluwalia, P. K., Electronic and optical properties of silicene under uni-axial and bi-axial mechanical strains: A first principle study. *Physica E: Low-dimensional Systems and Nanostructures*, 61, 2014.

59. Qin, R., Wang, C.-H., Zhu, W. & Zhang, Y., First-principles calculations of mechanical and electronic properties of silicene under strain. *AIP Advances*, 2, 2, 2012.

60. Xie, H. et al., Large tunability of lattice thermal conductivity of monolayer silicene via mechanical strain. *Physical Review B*, 93, 7, 2016.

61. Wang, S.-K. & Wang, J., Spin and valley filter in strain engineered silicene. *Chinese Physics B*, 24, 3, 2015.

62. Kaloni, T. P. & Schwingenschlögl, U., Stability of germanene under tensile strain. *Chemical Physics Letters*, 583, 2013.

63. Yarmohammadi, M., Electronic heat capacity and magnetic susceptibility of ferromagnetic silicene sheet under strain. *Solid State Communications*, 250, 2017.

64. Qin, R., Zhu, W., Zhang, Y. & Deng, X., Uniaxial strain-induced mechanical and electronic property modulation of silicene. *Nano Express*, 9, 2014.

65. Balendhran, S., Walia, S., Nili, H., Sriram, S. & Bhaskaran, M., Elemental analogues of graphene: silicene, germanene, stanene, and phosphorene. *Small*, 11, 6, 2015.

66. Broek, B.v.d. et al., Two-dimensional hexagonal tin: ab initio geometry, stability, electronic structure and functionalization. *2D Materials*, 1, 2, 2014.

67. Modarresi, M., Kakoee, A., Mogulkoc, Y. & Roknabadi, M. R., Effect of external strain on electronic structure of stanene. *Computational Materials Science*, 101, 2015.

68. Fu, B., Abid, M. & Liu, C.-C., Systematic study on stanene bulk states and the edge states of its zigzag nanoribbon. *New Journal of Physics*, 19, 10, 2017.

69. Xu, W. et al., Electronic and optical properties of Arsenene under uniaxial strain. *IEEE Journal of Selected Topics in Quantum Electronics*, 23, 1, 2017.

70. Kamal, C & Ezawa, M., Arsenene: Two-dimensional buckled and puckered honeycomb arsenic systems. *Physical Review B*, 91, 2015.

71. Wang, C., Xia, Q., Nie, Y., Rahman, M. & Guo, G., Strain engineering band gap, effective mass and anisotropic Dirac-like cone in monolayer arsenene. *AIP Advances*, 6, 3, 2016.

72. Zhang, S., Yan, Z., Li, Y., Chen, Z. & Zeng, H., Atomically thin arsenene and antimonene: semimetal-semiconductor and indirect-direct band-gap transitions. *Angew. Chem.*, 54, 10, 2015.

73. Ares, P., Palacios, J. J., Abellan, G., Gomez-Herrero, J. & Zamora, F., Recent progress on Antimonene: A new bidimensional material. *Advanced Materials*, 30, 2, 2018.

74. Fan, X. et al., Theoretical prediction of tunable electronic and magnetic properties of monolayer antimonene by vacancy and strain. *Applied Surface Science*, 488, 2019.

75. Zhao, M., Zhang, X. & Li, L., Strain-driven band inversion and topological aspects in Antimonene. *Scientific Reports*, 5, 2015.

76. Zhang, W. et al., Topological phase transitions driven by strain in monolayer tellurium. *Physical Review B*, 98, 11, 2018.

77. Ma, H., Hu, W. & Yang, J., Control of highly anisotropic electrical conductance of tellurene by strain-engineering. *Nanoscale*, 11, 45, 2019.

78. Cai, X., Ren, Y., Wu, M., Xu, D. & Luo, X., Strain-induced phase transition and giant piezoelectricity in monolayer tellurene. *Nanoscale*, 12, 1, 2020.

79. Zhu, Z., Cai, C., Niu, C., Wang, C., Sun, Q., Han, X., Guo, Z. & Jia, Y., Tellurene-a monolayer of tellurium from first-principles prediction. arXiv:1605.03253, 2016.

80. Singh, D., Gupta, S. K., Lukačević, I. & Sonvane, Y., Indiene 2D Monolayer: A new nanoelectronic material. *RSC Advances*, 6, 2016.

81. Muževič, M., Prediction and characterisation of low-dimensional structures of antimony, indium and aluminium[D]. University of Zagreb. Faculty of Science. Department of Physics, 2019.

82. Yeoh, K. H., Yoon, T. L., Rusi, Ong, D. S. & Lim, T. L., First-principles studies on the superconductivity of aluminene. *Applied Surface Science*, 445, 2018.

83. Drummond, N. D., Zólyomi, V. & Fal'ko, V. I., Electrically tunable band gap in silicene. *Physical Review B*, 85, 7, 2012.

84. Ni, Z. et al., Tunable bandgap in silicene and germanene. *Nano Letters*, 12, 1, 2012.

85. Jose, D. & Datta, A., Structures and electronic properties of silicene clusters: a promising material for FET and hydrogen storage. *Physical Chemistry Chemical Physics*, 13, 16, 2011.

86. Jose, D. & Datta, A., Understanding of the buckling distortions in Silicene. *The Journal of Physical Chemistry C*, 116, 46, 2012.

87. Jose Deepthi, D. A., Structures and chemical properties of Silicene: Unlike Graphene. *Accounts of Chemical Research*, 47, 2014.

88. Lou, P. & Lee, J. Y., Band structures of narrow Zigzag Silicon Carbon Nanoribbons. *The Journal of Physical Chemistry C*, 113, 2009.

89. Bishnoi, B. & Ghosh, B., Spin transport in silicene and germanene. *RSC Advances*, 3, 48, 2013.

90. Gürel, H. H., Özçelik, V. O. & Ciraci, S., Effects of charging and perpendicular electric field on the properties of silicene and germanene. *Journal of Physics: Condensed Matter*, 25, 2013.

91. Ni, Z. et al., Tunable band gap and doping type in silicene by surface adsorption: Towards tunneling transistors. *Nanoscale*, 6, 13, 2014.

92. Chowdhury, S. & Jana, D., A theoretical review on electronic, magnetic and optical properties of silicene. *Reports on Progress in Physics*, 79, 12, 2016.

93. Qin, G., Qin, Z., Yue, S. Y., Yan, Q. B. & Hu, M., External electric field driving the ultra-low thermal conductivity of silicene. *Nanoscale*, 9, 21, 2017.

94. Shu, H., Li, Y., Niu, X. & Guo, J., Electronic structures and optical properties of arsenene and antimonene under strain and an electric field. *Journal of Materials Chemistry C*, 6, 1, 2018.

95. Ghosal, S., Bandyopadhyay, A. & Jana, D., Electric field induced band tuning, optical and thermoelectric responses in tetragonal germanene: A theoretical approach. *Physical Chemistry Chemical Physics*, 22, 35, 2020.

96. Liu, M.-Y. et al., Strain and electric field tunable electronic structure of buckled bismuthene. *RSC Advances*, 7, 63, 2017.

97. Ghosh, B., Singh, B., Prasad, R. & Agarwal, A., Electric-field tunable Dirac semi-metal state in phosphorene thin films. *Physical Review B*, 94, 2016.

98. Yu, J., Katsnelson, M. I. & Yuan, S., Tunable electronic and magneto-optical properties of monolayer arsenene: From GW0 approximation to large-scale tight-binding propagation simulations. *Physical Review B*, 98, 11, 2018.

99. Le, P. T. T. & Yarmohammadi, M., Tuning thermoelectric transport in phosphorene through a perpendicular magnetic field. *Chemical Physics*, 519, 2019.

100. Romera, E., Roldán, J. B. & de los Santos, F., Zitterbewegung in monolayer silicene in a magnetic field. *Physics Letters A*, 378, 34, 2014.

Index

For Product Safety Concerns and Information please contact our EU
representative GPSR@taylorandfrancis.com
Taylor & Francis Verlag GmbH, Kaufingerstraße 24, 80331 München, Germany